# Data Analytics

# Systems Innovation Series

*Series Editor*
Adedeji B. Badiru
*Air Force Institute of Technology (AFIT) – Dayton, Ohio*

Systems innovation refers to all aspects of developing and deploying new technology, methodology, techniques, and best practices in advancing industrial production and economic development. This entails such topics as product design and development, entrepreneurship, global trade, environmental consciousness, operations and logistics, introduction and management of technology, collaborative system design, and product commercialization. Industrial innovation suggests breaking away from the traditional approaches to industrial production. It encourages the marriage of systems science, management principles, and technology implementation. Particular focus will be the impact of modern technology on industrial development and industrialization approaches, particularly for developing economics. The series will also cover how emerging technologies and entrepreneurship are essential for economic development and society advancement.

Productivity Theory for Industrial Engineering
*Ryspek Usubamatov*

Quality Management in Construction Projects
*Abdul Razzak Rumane*

Company Success in Manufacturing Organizations
A Holistic Systems Approach
*Ana M. Ferreras and Lesia L. Crumpton-Young*

Introduction to Industrial Engineering
*Avraham Shtub and Yuval Cohen*

Design for Profitability
Guidelines to Cost Effectively Manage the Development Process of Complex Products
*Salah Ahmed Mohamed Elmoselhy*

Data Analytics
Handbook of Formulas and Techniques
*Adedeji B. Badiru*

For more information about this series, please visit: https://www.crcpress.com/Systems-Innovation-Book-Series/book-series/CRCSYSINNOV

# Data Analytics
## Handbook of Formulas and Techniques

Adedeji B. Badiru

CRC Press is an imprint of the
Taylor & Francis Group, an **informa** business

First edition published 2020
by CRC Press
6000 Broken Sound Parkway NW, Suite 300, Boca Raton, FL 33487-2742

and by CRC Press
2 Park Square, Milton Park, Abingdon, Oxon, OX14 4RN

© 2021 Taylor & Francis Group, LLC

CRC Press is an imprint of Taylor & Francis Group, LLC

The right of Adedeji B. Badiru to be identified as author of this work has been asserted by him in accordance with sections 77 and 78 of the Copyright, Designs and Patents Act 1988.

Reasonable efforts have been made to publish reliable data and information, but the author and publisher cannot assume responsibility for the validity of all materials or the consequences of their use. The authors and publishers have attempted to trace the copyright holders of all material reproduced in this publication and apologize to copyright holders if permission to publish in this form has not been obtained. If any copyright material has not been acknowledged please write and let us know so we may rectify in any future reprint.

Except as permitted under U.S. Copyright Law, no part of this book may be reprinted, reproduced, transmitted, or utilized in any form by any electronic, mechanical, or other means, now known or hereafter invented, including photocopying, microfilming, and recording, or in any information storage or retrieval system, without written permission from the publishers.

For permission to photocopy or use material electronically from this work, access www.copyright.com or contact the Copyright Clearance Center, Inc. (CCC), 222 Rosewood Drive, Danvers, MA 01923, 978-750-8400. For works that are not available on CCC please contact mpkbookspermissions@tandf.co.uk

*Trademark notice*: Product or corporate names may be trademarks or registered trademarks and are used only for identification and explanation without intent to infringe.

*Library of Congress Cataloging-in-Publication Data*
Names: Badiru, Adedeji Bodunde, 1952- author.
Title: Data analytics : handbook of formulas and techniques / Adedeji Badiru.
Description: First edition. | Boca Raton, FL : CRC Press/Taylor & Francis
Group, LLC, 2021. | Series: Systems innovation book series |
Includes bibliographical references and index.
Identifiers: LCCN 2020033885 (print) | LCCN 2020033886 (ebook) |
ISBN 9780367537418 (hardback) | ISBN 9781003083146 (ebook)
Subjects: LCSH: Quantitative research—Handbooks, manuals, etc. |
Information visualization—Handbooks, manuals, etc. |
Engineering mathematics—Formulae—Handbooks, manuals, etc. |
Business mathematics—Handbooks, manuals, etc.
Classification: LCC QA76.9.Q36 B33 2021 (print) | LCC QA76.9.Q36 (ebook) |
DDC 001.4/2—dc23
LC record available at https://lccn.loc.gov/2020033885
LC ebook record available at https://lccn.loc.gov/2020033886

ISBN: 978-0-367-53741-8 (hbk)
ISBN: 978-1-003-08314-6 (ebk)

Typeset in Times
by codeMantra

*To the fond memories of the illuminating lectures and mentoring of Dr. Leland Long and Dr. Reginald Mazeres, my Mathematics professors at Tennessee Technological University, who heightened my interest in data visualization well before the name data analytics was coined.*

# Contents

Preface .................................................................................................................xxi
Acknowledgments ............................................................................................xxiii
Author .................................................................................................................xxv

**Chapter 1** Essentials of Data Analytics ............................................................ 1

    Introduction to COVID-19 Data Analytics ........................................ 1
    Systems View of Data Analytics ........................................................ 2
    Global Growth of Data Analytics ....................................................... 2
    Background in Predictive Analytics ................................................... 3
    Data Modeling Approaches ................................................................ 5
    Data Fanaticism .................................................................................. 6
    Data and Measurements for Data Analytics ...................................... 7
        What Is Measurement? ............................................................... 7
            Data Measurement Systems ............................................... 8
            Fundamental Scientific Equations .................................... 11
            Einstein's Equation ........................................................... 11
            Einstein's Field Equation ................................................. 11
            Heisenberg's Uncertainty Principle .................................. 12
            Schrödinger Equation ....................................................... 12
            Dirac Equation .................................................................. 12
            Maxwell's Equations ........................................................ 13
            Boltzmann's Equation for Entropy ................................... 13
            Planck–Einstein Equation ................................................. 13
            Planck's Blackbody Radiation Formula ........................... 14
            Hawking Equation for Black Hole Temperature ............. 14
            Navier–Stokes Equation for a Fluid ................................. 14
            Lagrangian for Quantum Chromodynamics .................... 14
            Bardeen–Cooper–Schrieffer Equation for Superconductivity .... 15
            Josephson Effect ............................................................... 15
            Fermat's Last Theorem .................................................... 15
        Methods for Data Measurement and Comparison ................... 16
            Direct Comparison ........................................................... 16
            Indirect Comparison ......................................................... 17
        Data Measurement Scales ........................................................ 17
            Nominal Scale of Measurement ....................................... 18
            Ordinal Scale of Measurement ......................................... 18
            Interval Scale of Measurement ......................................... 18
            Ratio Scale Measurement ................................................. 18
        Reference Units of Measurements ............................................ 19
        Common Constants .................................................................. 20
        Numeric Data Representation .................................................. 20

vii

|  |  |  |
|---|---|---|
|  | The Language of Data Analytics | 21 |
|  | Quick Reference for Mathematical Equations | 21 |
|  | Reference | 28 |
| **Chapter 2** | Empirical Model Building | 29 |
|  | Introduction to the Model Environment | 29 |
|  | State-Space Modeling | 30 |
|  | Calculus Reference for Data Analytics | 32 |
|  | Integration Rules | 33 |
|  | Solving Integrals with Variable Substitution | 34 |
|  | Riemann Integral | 34 |
|  | Integration by Parts | 34 |
|  |    Compound Functions Where the Inner Function Is $ax$ | 34 |
|  |    Integration by Parts | 35 |
|  | Systems Modeling for Data Analytics | 36 |
|  | Triple C Questions | 39 |
|  | Communication | 40 |
|  | Cooperation | 45 |
|  | Coordination | 47 |
|  | Conflict Resolution in Data Analytics | 47 |
|  | References | 49 |
| **Chapter 3** | Data Visualization Methods | 51 |
|  | Introduction to Data Visualization | 51 |
|  | Case Example of "Covidvisualizer" Website | 51 |
|  | Dynamism and Volatility of Data | 52 |
|  | Data Determination and Collection | 53 |
|  |    Choosing the Data | 53 |
|  |    Collecting the Data | 53 |
|  |    Relevance Check | 54 |
|  |    Limit Check | 54 |
|  |    Critical Value | 54 |
|  |    Coding the Data | 54 |
|  |    Processing the Data | 54 |
|  |    Control Total | 54 |
|  |    Consistency Check | 55 |
|  |    Scales of Measurement | 55 |
|  |    Using the Information | 55 |
|  |    Data Exploitation | 57 |
|  |       Raw Data | 57 |
|  |       Total Revenue | 58 |
|  |       Average Revenue | 59 |
|  |       Median Revenue | 61 |
|  |       Quartiles and Percentiles | 62 |

Contents　　　　　　　　　　　　　　　　　　　　　　　　　　　　ix

　　　　　　　　The Mode ................................................................................ 63
　　　　　　　　Range of Revenue .................................................................... 63
　　　　　　　　Average Deviation .................................................................... 63
　　　　　　　　Sample Variance ....................................................................... 64
　　　　　　　　Standard Deviation ................................................................ 65

**Chapter 4**　Basic Mathematical Calculations for Data Analytics ......... 69
　　　　　Introduction to Calculation for Data Analytics ............................... 69
　　　　　Quadratic Equation .......................................................................... 69
　　　　　　　　Overall Mean ........................................................................... 70
　　　　　　　　Chebyshev's Theorem ............................................................. 70
　　　　　　　　Permutations ............................................................................ 70
　　　　　　　　Combinations ........................................................................... 70
　　　　　　　　Failure ....................................................................................... 71
　　　　　　　　Probability Distribution .......................................................... 71
　　　　　　　　Probability ................................................................................ 71
　　　　　　　　Distribution Function ............................................................. 71
　　　　　　　　Expected Value ........................................................................ 72
　　　　　　　　Variance ..................................................................................... 72
　　　　　　　　Binomial Distribution .............................................................. 73
　　　　　　　　Poisson Distribution ................................................................ 73
　　　　　　　　Mean of a Binomial Distribution ........................................... 73
　　　　　　　　Variance ..................................................................................... 73
　　　　　　　　Normal Distribution ................................................................ 73
　　　　　　　　Cumulative Distribution Function ........................................ 73
　　　　　　　　Population Mean ..................................................................... 73
　　　　　　　　Standard Error of the Mean ................................................... 74
　　　　　　　　$t$-Distribution ........................................................................ 74
　　　　　　　　Chi-Squared Distribution ....................................................... 74
　　　　　Definition of Set and Notation ......................................................... 74
　　　　　Set Terms and Symbols .................................................................... 75
　　　　　Venn Diagrams ................................................................................... 75
　　　　　　　　Operations on Sets .................................................................. 76
　　　　　　　　De Morgan's Laws ................................................................... 76
　　　　　　　　Probability Terminology ......................................................... 77
　　　　　　　　Basic Probability Principles ................................................... 77
　　　　　　　　Random Variable ..................................................................... 77
　　　　　　　　Mean Value $\hat{x}$ or Expected Value $\mu$ ................... 78
　　　　　Series Expansions .............................................................................. 78
　　　　　Mathematical Signs and Symbols .................................................. 81
　　　　　Greek Alphabet .................................................................................. 83
　　　　　Algebra ................................................................................................ 83
　　　　　　　　Laws of Algebraic Operations ............................................... 83
　　　　　　　　Special Products and Factors ................................................. 83
　　　　　　　　Powers and Roots .................................................................... 85

- Proportion ............................................................................................. 85
- Arithmetic Mean of $n$ Quantities $A$ ............................................... 86
- Geometric Mean of $n$ Quantities $G$ ............................................... 86
- Harmonic Mean of $n$ Quantities $H$ ................................................ 86
- Generalized Mean ............................................................................... 86
- Solution of Quadratic Equations ....................................................... 87
- Solution of Cubic Equations .............................................................. 87
- Trigonometric Solution of the Cubic Equation ............................... 88
- Solution of Quadratic Equations ....................................................... 89
- Partial Fractions .................................................................................. 89
- Non-repeated Linear Factors ............................................................. 89
- Repeated Linear Factors .................................................................... 90
- General Terms ..................................................................................... 90
- Repeated Linear Factors .................................................................... 91
- Factors of Higher Degree .................................................................. 91
- Geometry ..................................................................................................... 91
  - Triangles ............................................................................................... 91
  - Right Triangle ...................................................................................... 92
  - Equilateral Triangle ............................................................................ 92
  - General Triangle ................................................................................. 92
    - Menelaus's Theorem ................................................................... 93
    - Ceva's Theorem ........................................................................... 93
  - Quadrilaterals ....................................................................................... 93
  - Rectangle .............................................................................................. 93
  - Parallelogram ....................................................................................... 94
  - Rhombus ............................................................................................... 94
  - Trapezoid .............................................................................................. 94
  - General Quadrilateral ......................................................................... 94
  - Regular Polygon of $n$ Sides Each of Length $b$ ............................ 95
  - Regular Polygon of $n$ Sides Inscribed in a Circle of Radius $r$ ....... 95
  - Regular Polygon of $n$ Sides Circumscribing a Circle of Radius $r$ .... 95
  - Cyclic Quadrilateral ............................................................................ 95
  - Prolemy's Theorem ............................................................................. 96
  - Cyclic-Inscriptable Quadrilateral ...................................................... 96
  - Planar Areas by Approximation ....................................................... 97
    - Trapezoidal Rule .......................................................................... 97
    - Durand's Rule ............................................................................... 97
    - Simpson's Rule ($n$ even) .......................................................... 97
    - Weddle's Rule ($n = 6$) .............................................................. 98
  - Solids Bounded by Planes ................................................................. 98
    - Cube ............................................................................................... 98
    - Rectangular Parallelepiped (or Box) ....................................... 98
  - Prism ...................................................................................................... 98
  - Pyramid ................................................................................................. 98
  - Prismatoid ............................................................................................ 99
  - Regular Polyhedra .............................................................................. 99

# Contents

Sphere of Radius $r$ .................................................................................... 100
Right Circular Cylinder of Radius $r$ and Height $h$ ...................... 100
Circular Cylinder of Radius $r$ and Slant Height $\ell$ ....................... 100
Cylinder of Cross-Sectional Area $A$ and Slant Height $\ell$ .............. 100
Right Circular Cone of Radius $r$ and Height $h$ ............................ 101
Spherical Cap of Radius $r$ and Height $h$ ...................................... 101
Frustum of Right Circular Cone of Radii $a$, $b$ and Height $h$ ....... 101
Zone and Segment of Two Bases .................................................. 101
Lune ............................................................................................... 101
Spherical Sector ............................................................................. 101
Spherical Triangle and Polygon .................................................... 101
Spheroids ....................................................................................... 102
    Ellipsoid ................................................................................ 102
    Oblate Spheroid ................................................................... 102
    Prolate Spheroid .................................................................. 102
    Circular Torus ...................................................................... 102
Formulas from Plane Analytic Geometry ........................................... 102
    Distance $d$ between Two Points ................................................. 102
    Slope $m$ of Line Joining Two Points .......................................... 103
    Equation of Line Joining Two Points ......................................... 103
    Equation of Line in Terms of $x$ Intercept $a \neq 0$ and $y$ intercept $b \neq 0$ ........................................................................................... 103
    Normal Form for Equation of Line ............................................ 103
    General Equation of Line ........................................................... 103
    Distance from Point $(x_1, y_1)$ to Line $Ax + By + C = 0$ ................. 103
    Angle $\psi$ between Two Lines Having Slopes $m_1$ and $m_2$ ............. 103
    Area of Triangle with Verticles .................................................. 104
    Transformation of Coordinates Involving Pure Translation ....... 104
    Transformation of Coordinates Involving Pure Rotation ........... 104
    Transformation of Coordinates Involving Translation and Rotation ...................................................................................... 104
    Polar Coordinates $(r, \theta)$ ............................................................. 105
    Plane Curves .............................................................................. 105
    Catenary, Hyperbolic Cosine ..................................................... 105
    Cardioid ..................................................................................... 105
    Circle ......................................................................................... 105
    Cassinian Curves ....................................................................... 105
    Cotangent Curve ........................................................................ 106
    Cubical Parabola ........................................................................ 106
    Cosecant Curve ......................................................................... 106
    Cosine Curve ............................................................................. 106
    Ellipse ........................................................................................ 106
    Gamma Function ....................................................................... 106
    Hyperbolic Functions ................................................................ 106
    Inverse Cosine Curve ................................................................ 107
    Inverse Sine Curve .................................................................... 107

Inverse Tangent Curve ..................................................................... 107
Logarithmic Curve ........................................................................... 107
Parabola ........................................................................................... 107
Cubical Parabola .............................................................................. 107
Tangent Curve .................................................................................. 107
Ellipsoid ........................................................................................... 107
Elliptic Cone .................................................................................... 107
Elliptic Cylinder ............................................................................... 107
Hyperboloid of One Sheet ............................................................... 108
Elliptic Paraboloid ............................................................................ 108
Hyperboloid of Two Sheets ............................................................. 108
Hyperbolic Paraboloid ..................................................................... 108
Sphere .............................................................................................. 108
Distance d between Two Points ...................................................... 108
Equations of Line Joining $P_1(x_1, y_1, z_1)$ and $P_2(x_2, y_2, z_2)$ in Standard Form ................................................................................. 108
Equations of Line Joining $P_1(x_1, y_1, z_1)$ and $P_2(x_2, y_2, z_2)$ in Parametric Form .............................................................................. 108
Angle $\phi$ between Two Lines with Direction Cosines $l_1, m_1, n_1$ and $l_2, m_2, n_2$ ............................................................................... 108
General Equation of a Plane ............................................................ 109
Equation of Plane Passing through Points ...................................... 109
Equation of Plane in Intercept Form ............................................... 109
Equations of Line through $(x_0, y_0, z_0)$ and Perpendicular to Plane ........................................................................................... 109
Distance from Point $(x, y, z)$ to Plane $Ax + By + D = 0$ ............. 109
Normal form for Equation of Plane ................................................. 109
Transformation of Coordinates Involving Pure Translation .......... 110
Transformation of Coordinates Involving Pure Rotation ............... 110
Transformation of Coordinates Involving Translation and Rotation ............................................................................................ 110
Cylindrical Coordinates $(r, \theta, z)$ ............................................... 111
Spherical Coordinates $(r, \theta, \phi)$ ............................................. 111
Logarithmic Identities ...................................................................... 112
Special Values .................................................................................. 112
Logarithms to General Base ............................................................ 112
Series Expansions ............................................................................ 113
Limiting Values ................................................................................ 113
Inequalities ....................................................................................... 113
Continued Fractions ......................................................................... 114
Polynomial Approximations ............................................................ 114
Fundamental Properties ................................................................... 115
Definition of General Powers .......................................................... 116
Logarithmic and Exponential Functions ......................................... 116
Polynomial Approximations ............................................................ 117
Slopes ............................................................................................... 125

Trigonometric Ratios .................................................................. 125
Sine Law .................................................................................... 127
Cosine Law ............................................................................... 127
Algebra ............................................................................................... 127
Expanding ................................................................................. 127
Factoring ................................................................................... 127
Roots of Quadratic ................................................................... 128
Law of Exponents .................................................................... 128
Logarithms ................................................................................ 128

**Chapter 5**  Statistical Methods for Data Analytics ........................................... 129

Introduction ....................................................................................... 129
Discrete Distributions ...................................................................... 129
Bernoulli Distribution .............................................................. 129
Beta Binomial Distribution ..................................................... 129
Beta Pascal Distribution .......................................................... 132
Binomial Distribution .............................................................. 132
Discrete Weibull Distribution ................................................. 132
Geometric Distribution ............................................................ 132
Hypergeometric Distribution .................................................. 133
Negative Binomial Distribution .............................................. 133
Poisson Distribution ................................................................ 134
Rectangular (Discrete Uniform) Distribution ....................... 134
Continuous Distributions ................................................................ 134
Arcsin Distribution .................................................................. 134
Beta Distribution ...................................................................... 135
Cauchy Distribution ................................................................. 135
Chi Distribution ....................................................................... 135
Chi-Square Distribution .......................................................... 135
Erlang Distribution .................................................................. 136
Exponential Distribution ......................................................... 136
Extreme-Value Distribution .................................................... 136
$F$ Distribution ........................................................................ 136
Gamma Distribution ................................................................ 137
Half-Normal Distribution ........................................................ 137
Laplace (Double Exponential) Distribution .......................... 138
Logistic Distribution ................................................................ 138
Lognormal Distribution ........................................................... 138
Noncentral Chi-Square Distribution ...................................... 139
Noncentral $F$ Distribution .................................................... 139
Noncentral $t$ Distribution ..................................................... 139
Normal Distribution ................................................................. 140
Pareto Distribution ................................................................... 140
Rayleigh Distribution .............................................................. 140
$t$ Distribution .......................................................................... 141

- Triangular Distribution .................................................................. 141
- Uniform Distribution ..................................................................... 142
- Weibull Distribution ...................................................................... 142
- Distribution Parameters ................................................................. 142
  - Average ....................................................................................... 142
  - Variance ...................................................................................... 142
  - Standard Deviation ..................................................................... 143
  - Standard Error ............................................................................ 143
  - Skewness .................................................................................... 143
  - Standardized Skewness .............................................................. 143
  - Kurtosis ....................................................................................... 143
  - Standardized Kurtosis ................................................................. 143
  - Weighted Average ...................................................................... 143
- Estimation and Testing .................................................................... 144
  - $100(1-\alpha)\%$ Confidence Interval for Mean ............................... 144
  - $100(1-\alpha)\%$ Confidence Interval for Variance ........................... 144
  - $100(1-\alpha)\%$ Confidence Interval for Difference in Means ......... 144
    - Equal Variance ....................................................................... 144
    - Unequal Variance ................................................................... 144
  - $100(1-\alpha)\%$ Confidence Interval for Ratio of Variances ............ 145
  - Normal Probability Plot .............................................................. 145
  - Comparison of Poisson Rates ...................................................... 145
- Distribution Functions and Parameter Estimation ........................... 146
  - Bernoulli ..................................................................................... 146
  - Binomial ..................................................................................... 146
  - Discrete Uniform ........................................................................ 146
  - Geometric ................................................................................... 146
  - Negative Binomial ...................................................................... 146
  - Poisson ....................................................................................... 146
  - Beta ............................................................................................ 146
  - Chi-Square .................................................................................. 147
  - Erlang ......................................................................................... 147
  - Exponential ................................................................................ 147
  - $F$ ................................................................................................ 147
  - Gamma ....................................................................................... 147
  - Lognormal .................................................................................. 148
  - System Displays ......................................................................... 148
  - Normal ....................................................................................... 148
  - Student's $t$ ................................................................................. 148
  - Triangular ................................................................................... 148
  - Uniform ...................................................................................... 149
  - Weibull ....................................................................................... 149
  - Chi-Square Test for Distribution Fitting ..................................... 149
  - Kolmogorov–Smirnov Test .......................................................... 149
  - ANOVA (Analysis of Variance) .................................................. 150
    - Notations ................................................................................ 150

## Contents

- Standard Error (Internal) ..... 151
- Standard Error (Pooled) ..... 151
- Interval Estimates ..... 151
- Tukey Interval ..... 151
- Scheffe Interval ..... 151
- Cochran $C$-Test ..... 151
- Bartlett Test ..... 152
- Hartley's Test ..... 152
- Kruskal–Wallis Test ..... 152
- Freidman Test ..... 153
- Regression ..... 154
  - Notations ..... 154
  - Regression Statistics ..... 154
  - Predictions ..... 155
  - Nonlinear Regression ..... 156
  - Ridge Regression ..... 157
  - Quality Control ..... 157
  - Subgroup Statistics ..... 157
  - $X$-Bar Charts ..... 158
  - Capability Ratios ..... 159
  - $R$ Charts ..... 159
  - $S$ Charts ..... 160
  - $C$ Charts ..... 160
  - $U$ Charts ..... 160
  - $P$ Charts ..... 160
  - $NP$ Charts ..... 161
  - CuSum Chart for the Mean ..... 161
  - Multivariate Control Charts ..... 162
- Time Series Analysis ..... 162
  - Notations ..... 162
  - Autocorrelation at Lag $k$ ..... 162
  - Partial Autocorrelation at Lag $k$ ..... 163
  - Cross-Correlation at Lag $k$ ..... 163
  - Box-Cox ..... 163
  - Periodogram (Computed Using Fast Fourier Transform) ..... 164
- Categorical Analysis ..... 164
  - Notations ..... 164
  - Totals ..... 164
  - Chi-Square ..... 165
  - Fisher's Exact Test ..... 165
  - Lambda ..... 165
  - Uncertainty Coefficient ..... 165
  - Somer's $D$ ..... 166
  - Eta ..... 167
  - Contingency Coefficient ..... 167
  - Cramer's $V$ ..... 168

  Conditional Gamma................................................................. 168
  Pearson's $r$............................................................................ 168
  Kendall's Tau $b$ .................................................................... 168
  Tau $C$..................................................................................... 168
 Probability Terminology............................................................ 168
 Basic Probability Principles ..................................................... 169
 Random Variable ....................................................................... 169
 Mean Value $\hat{x}$ or Expected Value $\mu$ ........................................ 169
Discrete Distribution Formulas ....................................................... 170
 Bernoulli Distribution ............................................................... 170
 Beta Binomial Distribution ...................................................... 170
 Beta Pascal Distribution ........................................................... 171
 Binomial Distribution................................................................ 171
 Discrete Weibull Distribution................................................... 171
 Geometric Distribution.............................................................. 171
 Hypergeometric Distribution.................................................... 171
 Negative Binomial Distribution................................................ 172
 Poisson Distribution ................................................................. 172
 Rectangular (Discrete Uniform) Distribution ........................... 173
 Continuous Distribution Formulas ........................................... 173
 Arcsin Distribution ................................................................... 173
 Beta Distribution ....................................................................... 173
 Cauchy Distribution................................................................... 174
 Chi Distribution......................................................................... 174
 Chi-Square Distribution ............................................................ 174
 Erlang Distribution .................................................................... 174
 Exponential Distribution ........................................................... 175
 Extreme-Value Distribution ...................................................... 175
 F Distribution............................................................................. 175
 Gamma Distribution .................................................................. 176
 Half-Normal Distribution .......................................................... 176
 Laplace (Double Exponential) Distribution ............................. 176
 Logistic Distribution................................................................. 177
 Lognormal Distribution............................................................. 177
 Noncentral Chi-Square Distribution ......................................... 177
 Noncentral $F$ Distribution ....................................................... 178
 Noncentral $t$ Distribution ........................................................ 178
 Normal Distribution .................................................................. 178
 Pareto Distribution .................................................................... 179
 Rayleigh Distribution................................................................ 179
 $t$ Distribution........................................................................... 179
 Triangular Distribution.............................................................. 180
 Uniform Distribution................................................................. 180
 Weibull Distribution.................................................................. 180
 Variate Generation Techniques ................................................. 181
Reference........................................................................................... 183

| | | |
|---|---|---|
| **Chapter 6** | Descriptive Statistics for Data Presentation | 185 |
| | Introduction | 185 |
| |    Sample Average | 185 |
| |    Sample Variance | 185 |
| |    Sample Standard Deviation | 186 |
| |    Sample Standard Error of the Mean | 187 |
| |       Skewness | 187 |
| |       Standardized Skewness | 188 |
| |       Kurtosis | 188 |
| |       Standardized Kurtosis | 188 |
| |       Weighted Average | 188 |
| | Estimation and Testing | 188 |
| |    $100(1-\alpha)\%$ Confidence Interval for Mean | 188 |
| |    $100(1-\alpha)\%$ Confidence Interval for Variance | 188 |
| |    $100(1-\alpha)\%$ Confidence Interval for Difference in Means | 188 |
| |       For Equal Variance | 188 |
| |       For Unequal Variance | 189 |
| |    $100(1-\alpha)\%$ Confidence Interval for Ratio of Variances | 189 |
| |    Normal Probability Plot | 189 |
| |    Comparison of Poisson Rates | 190 |
| | Distribution functions and Parameter Estimation | 190 |
| |    Bernoulli Distribution | 190 |
| |    Binomial Distribution | 190 |
| |    Discrete Uniform Distribution | 190 |
| |    Geometric Distribution | 190 |
| |    Negative Binomial Distribution | 191 |
| |    Poisson Distribution | 191 |
| |    Beta Distribution | 191 |
| |    Chi-Square Distribution | 191 |
| |    Erlang Distribution | 191 |
| |    Exponential Distribution | 192 |
| |    F Distribution | 192 |
| |    Gamma Distribution | 192 |
| |    Lognormal Distribution | 192 |
| |    Normal Distribution | 193 |
| |    Student's t | 193 |
| |    Triangular Distribution | 193 |
| |    Uniform Distribution | 193 |
| |    Weibull Distribution | 193 |
| |    Chi-Square Test for Distribution Fitting | 194 |
| |    Kolmogorov–Smirnov Test | 194 |
| |    ANOVA (Analysis of Variance) | 194 |
| |       Notations | 194 |
| |       Standard Error | 195 |
| |       Interval Estimates | 195 |

| | |
|---|---|
| Tukey Interval | 195 |
| Scheffe Interval | 196 |
| Cochran C-test | 196 |
| Bartlett Test | 196 |
| Hartley's Test | 196 |
| Kruskal–Wallis Test | 197 |
| Freidman Test | 197 |
| Regression | 198 |
|     Notations | 198 |
|     Statistical Quality Control | 199 |
|     Subgroup Statistics | 199 |
|     $X$-Bar Charts | 199 |
|     Capability Ratios | 200 |
|     $R$ Charts | 201 |
|     S Charts | 201 |
|     $C$ Charts | 201 |
|     $U$ Charts | 201 |
|     $P$ Charts | 201 |
|     $NP$ Charts | 202 |
|     CuSum Chart for the Mean | 202 |
| Time Series Analysis | 203 |
|     Notations | 203 |
|     Autocorrelation at Lag $k$ | 203 |
|     Partial Autocorrelation at Lag $k$ | 203 |
|     Cross-Correlation at Lag $k$ | 203 |
|     Box-Cox Computation | 204 |
|     Periodogram (Computed Using Fast Fourier Transform) | 204 |
| Categorical Analysis | 205 |
|     Notations | 205 |
|     Totals | 205 |
|     Chi-Square | 205 |
|     Lambda | 205 |
|     Uncertainty Coefficient | 206 |
|     Somer's $D$ Measure | 206 |
|     Eta | 207 |
|     Contingency Coefficient | 208 |
|     Cramer's V Measure | 208 |
|     Conditional Gamma | 208 |
|     Pearson's $r$ Measure | 208 |
|     Kendall's Tau $b$ Measure | 208 |
|     Tau C Measure | 209 |
|     Overall Mean | 209 |
|     Chebyshev's Theorem | 209 |
|     Permutation | 209 |
|     Combination | 209 |
|     Failure | 209 |

**Chapter 7**   Data Analytics Tools for Understanding Random Field
              Regression Models ............................................................................. 211

              Introduction .................................................................................... 211
              RFR Models .................................................................................... 212
              Two Examples ................................................................................ 214
              Bayesian Regression Models and Random Fields ............................ 214
              Data Analysis: Finding the Associated Regression Model .............. 215
              Relating Eigenvectors to Regression Functions ................................ 223
              Some Special Random Field Models ............................................... 225
              Gaussian Covariance as Damped Polynomial Regression ................ 225
              Trigonometric Regression and Spline Covariance ........................... 227
              Discussion....................................................................................... 228
              References ...................................................................................... 229

**Chapter 8**   Application of DEJI Systems Model to Data Integration ................. 233

              Introduction to Data Integration ..................................................... 233
              Leveraging the Input-Control-Output-Mechanism Model ............... 234
              Data Types and Fidelity .................................................................. 235
              Data Collection and Sanitization ..................................................... 236
              DEJI Systems Model for Data Quality ............................................. 238
              Data Value Model ........................................................................... 239
              Data Quality Control ...................................................................... 241
              References ...................................................................................... 242

**Index** ........................................................................................................... 243

# Preface

The year 2020 represents a particularly challenging year for humanity because of the COVID-19 pandemic caused by the coronavirus. No part of the entire world was spared from the ravages of the pandemic. As the world scrambled around to find a response, treatment, or cure, data was at the core of all efforts. All types of data analytics initiatives emerged to try and make sense of what was happening. The fast-spread pace of COVID-19 necessitated new and renewed interest in the tools and techniques of data collection, analysis, presentation, and sense-making. The unanticipated developments provided the motivation for writing this book. Having lived my industrial engineering profession within the hierarchical realm of data, information, and decision, I embarked on the arduous task of preparing this manuscript. Fast and good decisions were needed. This meant that data analytics was urgently needed. Good data analytics is the basis for effective decisions. Whoever has the data has the ability to extract information promptly and effectively to make pertinent decisions. Good data is useless if we cannot extract useful information relevant for the decision at hand. The ability to extract useful information is enhanced by data analytics tools. The premise of this book is to empower users and tool developers with the appropriate collection of formulas and techniques for data analytics. This will serve as a quick reference to keep pertinent formulas within fingertip reach of readers. It is well understood that we cannot manage or control anything that we cannot measure. Data analytics is the cornerstone of achieving pertinent measurement for the control of that which needs to be controlled. Therein lies the value of this book. The formulas and techniques contained in this book will be useful and relevant for all categories of readers. Its focus and conciseness will make the book appealing. Contents include formulas for descriptive statistics, estimation, testing, parametric analysis, distribution functions, analysis of variance, regression, quality control, time series analysis, and categorical data analysis.

**Wishing readers fast data responses,**
**Adedeji B. Badiru**
*July 2020*

# Acknowledgments

Once again, I acknowledge the consistent and professional support of Mrs. Cindy Renee Carelli and her colleagues at CRC Press/Taylor & Francis Group for the production of this book. The expert handling of the entire process made it possible to get the project done expeditiously with particular attention to all the pertinent details. Ms. Erin Harris is also recognized for her superior execution of all the administrative functions related to all phases of the book's production. Much of the typing and word processing of this manuscript was a family affair. My wife, Iswat, and my son, TJ, helped to type many of the numerous complex equations contained in the manuscript. My great appreciation and thanks go to both of them.

# Author

**Adedeji B. Badiru** is professor of Systems Engineering in the Graduate School of Engineering and Management at the Air Force Institute of Technology. He was previously professor and department head of Industrial Engineering at the University of Tennessee and professor of Industrial Engineering at the University of Oklahoma. He is a registered professional engineer (PE), a certified project management professional (PMP), and a fellow of the Institute of Industrial & Systems Engineers. He is the author of more than 30 books, 34 book chapters, and 80 technical journal articles. He is a member of several professional associations and scholastic honor societies.

# 1 Essentials of Data Analytics

Data, data, data everywhere, but not a bit to byte and chew on.

## INTRODUCTION TO COVID-19 DATA ANALYTICS

This equation-and-formula-based book is the result of COVID-19-driven interest in data analytics. The worldwide pandemic of COVID-19, the disease caused by the novel coronavirus, in early 2020 heightened the awareness and interest in data analytics. Prior to this particular pandemic, most people paid little attention to the efficacy of data analytics. As the horror of the pandemic unfolded, the primary connectivity that citizens had to the latest gory developments was through output of data analytics presented in streaming updates of visualization charts, tables, and numeric details. In fact, the hurried global assembling of COVID-19 data served as the initial motivation that initiated my foray into writing this handbook. The increased interest in data analytics is for the purpose of visualizing and predicting the spread of the disease. Thus, the need for this handbook is predicted on the growing demand for additional references on data analytics. The handbook focuses on providing the mathematical backbone upon which data analytics tools and software can be developed. As such, the premise of the handbook lies on data analytics formulas and techniques. Students and researchers developing data models can use the contents as guiding references for what is available in terms of modeling techniques. Data analytics can be used for data diagnostics for COVID-19 and other pandemics. Data analytics is needed for a prove of concept of new developments or new treatments protocols. Data analytics can help to achieve the following:

- Know the data we have.
- Know what the data says.
- Know how to analyze the data.
- Know how to present the data for relevant decisions.

The ubiquitous Microsoft Excel has powerful tools that are based on formulas and equations of the genre contained in this book. The spreadsheet software can do database queries, illustrative plots, simulation, optimization, machine learning, and curve fitting. Creative and experienced users can develop customized uses of the software. Coronavirus modeling requires formulas, equations, and techniques similar to the contents of this book.

## SYSTEMS VIEW OF DATA ANALYTICS

Coronavirus has disrupted the entire world system. No place is immune because there is no previous immunity against the virus. This brings us to take a systems view of the world. In systems engineering, a system is defined as follows:

> A collection of interrelated elements, whose combined output is higher than the sum of the <u>individual</u> outputs of the elements making up the system.

I offer the following COVID-19 definition of a contagion-prone world system:
COVID-19 Novel Definition of a World System

> A COVID-19 World System is a group of inter-connected regions of the world, whose normally-localized individual contagion has spread to infect the entire World.

Flattening the COVID-19 curve has been the calling card of governments and organizations. But we must know what the curve looks like before we can flatten it. To know the curve, we must know the data. To see the power of the data, we must use data analytics. To use data analytics, we must understand the underlying formulas and techniques. Therein lies the efficacy of this handbook.

## GLOBAL GROWTH OF DATA ANALYTICS

> You don't go to war with the data you need. You go to war with the data you have.
>
> *Dilbert Comic, May 7, 2020*

Even before the advent of COVID-19, "data," in all its ramifications, is of interest in every organization, whether in business, industry, government, the military, or academia. Research and practical applications of data have led to alternate names for what we do with data or how we process it. Consequently, we have seen alternate entries in what this author refers to as "Data Terminology Dictionary," which contains names such as the ones in the following list:

- Data science
- Data analytics
- Big data
- Data mining
- Data awareness
- Data familiarization
- Data socializing
- Data visualizing
- Data envelopment
- Data minding
- Data processing
- Data analysis
- Data management
- Data modeling

Over the past several decades, there have been significant developments in the application of quantitative techniques for decision-making in most areas of human activities. The success of such applications, however, depends on the quality of information available to the decision maker. That essential information is derived from archival or live data. Analysts have always struggled with developing new tools and techniques to extract useful information from the available data. The necessary information often involves future developments, which are, in most cases, nondeterministic. In such unfortunate situations, forecasting and projections are the viable means of obtaining the much-desired information. The problem then is to formulate methodologies, tools, techniques, and models that could efficiently and consistently yield reliable forecasts, projections, and trend lines. The search for enhanced approaches has led to the emergence of data-oriented movements, such as Data Mining, Big Data, and Data Envelopment. More recently, the previous approaches have morphed into the hot areas of data analytics, data science, data visualization, and data modeling. These new names are like new wine in an old bottle. What is different now is that emergence of new powerful computational tools, such as supercomputers, neural networks, and artificial intelligence, that make it possible to quickly analyze large volumes of data.

Interest in data analytics is spreading rapidly all over the world. This fact is the reason that most universities are now starting new degree programs and certificate offerings in data analytics, data science, and data modeling. This book is directly applicable to this evidence. The language of mathematics is universal. Formulas, equations, and quantitative modeling are the basis for data analytics. This handbook is designed as a useful reference for the tools and techniques of data analytics. For the quantitative-inclined readers, the formulas will be appealing. Illustrative examples will increase the comfort level with how each technique is used, thus helping to increase the use of data analytics to improve decision-making in organizations. This handbook can be a ready reference to what is contained in data analytics software tools. Analytics is presently the rage in business, industry, government, and academia. The basis of all operations in any environment rests on a foundation of data analytics. This handbook provides additional incentive to embrace data analytics and mitigate the proverbial risk of "garbage in, garbage out."

## BACKGROUND IN PREDICTIVE ANALYTICS

My own personal interest in data analytics dates back to my high school days at Saint Finbarr's College, Lagos, Nigeria, in the 1960s. Interestingly, that interest was spurred by the historical accounts of how Florence Nightingale used statistics as a tool for promoting a crusade for hygiene throughout the population. One of our teachers at Saint Finbarr's College was particularly enamored with Florence's historical exploits, and he lectured repeatedly about her accomplishments to the extent that I developed an interest in how statistics could be used for personal planning even at that formative stage of my education. Florence Nightingale's data-driven practices were credited with significantly improving hospital efficiency of that era. I was very intrigued.

Florence Nightingale was a British social reformer and statistician, and the founder of modern nursing. Nightingale came to prominence while serving as a manager and trainer of nurses during the Crimean War (October 1853–February 1856), in which she organized care for wounded soldiers.

With an interest in boosting my academic performance in selected subjects, I started developing hand-drawn charts to do trend analysis of my course grades. It was an interesting application that caught the attention of my classmates and teachers. With the trend charts, I could see the fluctuations in my course grades, based on which I intensified by study habits either to maintain or improve outcomes in specific courses. It was a rudimentary application of visual data analytics for a personal topic of interest, although the name "data analytics" was far from what I would have called it. It worked. I did not think much of that basal tool of course-grade statistics until I started my industrial engineering studies at Tennessee Technological University many years later. As of today, I still have the original hand-drawn data analytics trend chart of my high school course grades from 1968 through 1972. Although faded in resolution and finessing, it still conveys my long-standing interest and dogged pursuit of data analytics. Figure 1.1 presents an archival illustrative rendering of what the 1972 chart looks like. It is like taking a step back in data visualizing time. It is important to re-emphasize that Figure 1.1 is a scanned reprint of the original hand-drawn chart from 1968 to 1972. The importance of preserving the original appearance of the chart, albeit in low-resolution print quality, is to convey that data visualization does not have to be esoteric. Even a rudimentary hastily hand-drawn chart may provide sufficient information for corrective or proactive actions to enhance a desired end result. The chart in Figure 1.1, as poor in print quality as it may appear here, did provide the desired impact over the grade-tracking years of 1968 through 1972.

Along the line of my high school course performance trend charts, at Tennessee Technological University graduate school, I did a 1981 master's thesis research on stochastic modeling of energy consumption at the university (Badiru, 1981). The thesis research concerned the development of a stochastic forecasting model for energy consumption at the university. Such a model is useful for obtaining reliable forecasts of future energy consumption. This was an important pursuit in that era because of

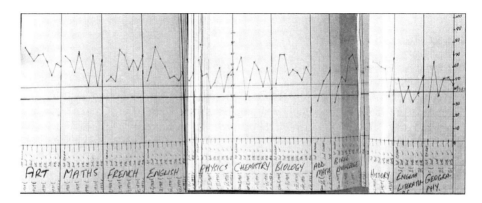

**FIGURE 1.1** High school grade data analytics: Academic performance trend line.

# Essentials of Data Analytics

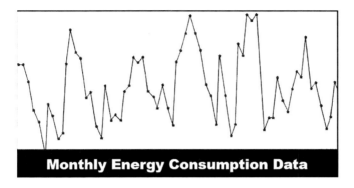

**FIGURE 1.2** Time series sample data for energy consumption data analytics.

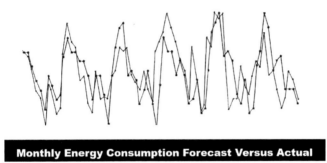

**FIGURE 1.3** Energy consumption forecast versus actual comparison.

the energy crisis of the 1979–1980 period. Tennessee Tech energy forecasts were important for managerial decisions, short-range planning, conservation efforts, and budget preparations. Using the ARIMA (Autoregressive Integrated Moving Average) time series modeling approach (Badiru, 1981), I constructed a forecast model, using a set of six-year data obtained from the university's physical plan office. A computer program was written in FORTRAN programming language to achieve the model building and the forecasting processes. Figures 1.2 and 1.3 illustrate samples of the plotting achieved from the time series model.

## DATA MODELING APPROACHES

Several models and variations for data analytics are used in practice. Below are some of the more common models:

1. Conceptual models
2. Theoretical models
3. Observational models
4. Experimental models

5. Empirical models
6. Iconic models
7. A priori reasoning

The conceptual data model is a structured business view of the data required to support business processes, record business events, and track-related performance measures. This model focuses on identifying the data used in the business but not its processing flow or physical characteristics.

So, a theoretical model can be defined as a theory that is developed to explain a situation or a phenomenon and, further, to be able to forecast it. Theoretical modeling is based on a number or a set of theories. These theories are used to explain some situations, phenomena, and behavior types.

This term is more commonly used in qualitative research, while the term "theoretical model" usually appears as a tool in quantitative research. They both refer to the key theories, models, and ideas that exist in relation to your chosen topic. They give your research a direction and set boundaries for the reader.

Observational learning is a type of learning that happens indirectly through a process of watching others and then imitating, or modeling, their behavior, with the imitating being called modeling. His 1961 Bobo doll experiment demonstrated how school-aged children modeled aggressive behavior seen in adults.

Experimental model is an example of conducting experiments to collect data to develop understanding that can then be transferred to another situation of interest. A common example is the use of animal experiments to model the human case. Animals that are employed to model the development and progression of diseases, and to test new treatments before they are given to humans, are used in experimental modeling. Animals with transplanted human cancers or other tissues are called xenograft models.

Empirical modeling refers to any kind of (computer) modeling based on empirical observations rather than on mathematically describable relationships of the system modeled. Empirical modeling is a generic term for activities that create models by observation and experiment. Empirical Modeling (with the initial letters capitalized, and often abbreviated to EM) refers to a specific variety of empirical modeling in which models are constructed following particular principles.

An iconic model is an exact physical representation and may be larger or smaller than what it represents. The characteristics of an iconic model and the object that it represents are the same. This model is frequently used in Operations Research.

In a priori reasoning, the modeling is based on knowledge or inclination originating from theoretical deduction rather than from observation, experience, or empirical data assessment.

## DATA FANATICISM

As has been illustrated with the examples presented in this chapter, the passion and reliance for data suggests the growth of data fanaticism, whereby people demand data as the basis for believing whatever is presented to them. This became more obvious during public briefings on COVID-19 pandemic in the early to mid-2020s.

Several newspaper, radio, and television programs bombarded the airwaves with segments on "Data and Facts" and "Facts and Fears" of COVID-19. The phrase "show us the data and we will believe" became a frequent call from the public as government officials scramble to convince the public to maintain social distancing and wear face masks to curtail the spread of coronavirus. These public observations heightened my interest in writing this book. Having used data analytics to drive personal academic performance and using data analytics on official job functions, my fervor to compile a guide for others persisted throughout the manuscript-writing chore.

## DATA AND MEASUREMENTS FOR DATA ANALYTICS

Measurement pervades everything we do. This applies to technical, management, and social activities and requirements. Even in innocuous situations, such as human leisure, the importance of data and measurement comes to the surface. How much, how far, how good, how fast, how long, how high, how many, and how often are typical conveyances of some sort of measurement. This has taken on an even more significant implication under the ravages of the COVID-19 pandemic, where there is daily tracking of the number of virus tests the number of infections, the number of hospitalizations, and the number of deaths. It is on the basis of data, measurements, and tracking that we can aspire to straighten or bend the curve of COVID-19 pandemic or other pandemics of the future.

### What Is Measurement?

It is well understood that we cannot manage anything if we cannot measure it. All elements involved in our day-to-day decision-making involve some form of measurement. Measuring an attribute of a system and then analyzing it against some standard, some best practice, or some benchmark empowers a decision maker to take appropriate and timely actions.

Fundamentally, measurement is the act or the result of a quantitative comparison between a predefined standard and an unknown magnitude. If the result is to be generally meaningful, two requirements must be met in the act of measurement:

1. The standard that is used for comparison must be accurately known and commonly accepted.
2. The procedure and instrument employed for obtaining the comparison must be provable and repeatable.

The first requirement is that there is an accepted standard of comparison. A weight cannot simply be heavy. It can only be proportionately as heavy as something else, namely, the standard. A comparison must be made, and unless it is made relative to something generally recognized as a standard, the measurement can only have a limited meaning. This holds for any quantitative measurement we may wish to make. In general, the comparison is one of magnitude, and a numerical result is presupposed. The quantity in question may be twice as large as the standard, or 1.5 times as large, or in some other ratio, but a numerical comparison must be made for it to be meaningful. The typical characteristics of a measurement process include the following:

- Precision
- Accuracy
- Correlation
- Stability
- Linearity
- Type of data

## Data Measurement Systems

The worldwide spread of COVID-19 has further heightened the recognition that different systems of measurement exist. The two primary and most commonly used systems are the **English system** and the **metric system**. For a good comparative assessment of data collected across the world, we need to be more cognizant of the differences and similarities between the two systems and the conversion relationships. For example, body temperature is one of the primary symptoms assessed in the initial diagnosis of COVID-19. In the English system, the standard body temperature is 98.6°F. Anything above 101.4°F is adjudged to represent a fever, a key signal of an infection. The question, often, is why are those two temperature numbers not whole numbers? Why don't we use 98°F, 99°F, 101°F, or 102°F? The temperature points are actually whole numbers in the metric system, where the normal body temperature is 37°C and a fever is indicated at 38°C. It is upon converting the 37°C and 38°C to Fahrenheit that we get 98.6°F and 101.4°F, respectively. Thus, it is important to have an appreciation of the different measurement systems and their conversion relationships.

The English system is the system that is commonly used in the United States today, whereas the metric system is used in many other parts of the world. The American measurement system is nearly the same as that brought by the American colony settlers from England. These measures had their origins in a variety of cultures, including Babylonian, Egyptian, Roman, Anglo-Saxon, and Nordic French. The ancient "digit," "palm," "span," and "cubic" units of length slowly lost preference to the length units "inch," "foot," and "yard." Roman contributions include the use of 12 as a base number and the words from which we derive many of the modern names of measurement units. For example, the 12 divisions of the Roman "pes" or foot were called unciae. The "foot" as a unit of measuring length is divided into 12 inches. The common words "inch" and "ounce" are both derived from Latin words. The "yard" as a measure of length can be traced back to early Saxon kings. They wore a sash or girdle around the waist that could be removed and used as a convenient measuring device. Thus, the word "yard" comes from the Saxon word "gird," which represents the circumference of a person's waist, preferably as "standard person," such as a king.

Evolution and standardization of measurement units often had interesting origins. For example, it was recorded that King Henry I decreed that a yard should be the distance from the tip of his nose to the end of his outstretched thumb. The length of a furlong (or furrow-long) was established by early Tudor rulers as 220 yards. This led Queen Elizabeth I to declare in the 16th century that the traditional Roman mile of 5,000 feet would be replaced by one of 5,280 feet, making the mile exactly eight furlongs and providing a convenient relationship between the furlong and the mile. To this day, there are 5,280 feet in one mile, which is 1,760 yards. Thus, through royal edicts, England, by the 18th century, had achieved a greater degree of standardization than other European countries. The English units were well suited to commerce and trade because they had been developed and refined to meet commercial needs. Through English colonization and its dominance of world commerce during the 17th, 18th, and 19th centuries, the English system of measurement units became established in many parts of the world, including the American colonies. The early 13 American colonies, however, had undesirable differences with respect to measurement standards for commerce. The need for a greater uniformity led to clauses in the Articles of Confederation (ratified by the original colonies in 1781) and the Constitution of the United States (ratified in 1788) that gave Congress the power to fix uniform standards for weights and measures across the colonies. Today, standards provided by the U.S. National Institute of Standards and Technology (NIST) ensure uniformity of measurement units throughout the country.

The need for a single worldwide coordinated measurement system was recognized over 300 years ago. In 1670, Gabriel Mouton, Vicar of St. Paul's Church in Lyons and an astronomer, proposed a comprehensive decimal measurement system based on the length of one minute of arc of a great circle of the Earth. Mouton also proposed the swing length of a pendulum with a frequency of one beat per second as the unit of length. A pendulum with this beat would have been fairly easily reproducible, thus facilitating the widespread distribution of uniform standards.

In 1790, in the midst of the French Revolution, the National Assembly of France requested the French Academy of Sciences to "deduce an invariable standard for all the measures and all the weights." The Commission appointed by the Academy created a system that was, at once, simple and scientific. The unit of length was to be a portion of the Earth's circumference. Measures for capacity (volume) and mass were to be derived from the unit of length, thus relating the basic units of the system to each other and to nature. Furthermore, larger and smaller multiples of each unit were to be created by multiplying or dividing the basic units by 10 and powers of 10. This feature provided a great convenience to users of the system, by eliminating the need for such calculations as dividing by 16 (to convert ounces to pounds) or by 12 (to convert inches to feet). Similar calculations in the metric system could be performed simply by shifting the decimal point. Thus, the metric system is a "base-10" or "decimal" system.

The Commission assigned the name metre (i.e., meter in English) to the unit of length. This name was derived from the Greek word *metron*, meaning "a measure." The physical standard representing the meter was to be constructed so that it would equal one ten-millionth of the distance from the North Pole to the equator along the meridian running near Dunkirk in France and Barcelona in Spain. The initial metric unit of mass, the "gram," was defined as the mass of one cubic centimeter (a cube that is 0.01 m on each side) of water at its temperature of maximum density. The cubic decimeter (a cube 0.1 m on each side) was chosen as the unit for capacity. The fluid volume measurement for the cubic decimeter was given the name "liter." Although the metric system was not accepted with much enthusiasm at first, adoption by other nations occurred steadily after France made its use compulsory in 1840. In my own personal educational case, I grew up in Nigeria in the 1950s, 1960s, and early 1970s, under the English system of measurement. Nigeria converted to the metric system in 1975, the year that I proceeded to the United States for further studies. So, I never underwent the necessity to educationally convert to the metric system since the United States continued to use the English system. Children born in Nigeria after the conversion went through the metric system of measurement in school. Attempts by the United States to convert to the metric system in the late 1970s and early 1980s never really took hold. For me today, I still straddle both systems of measurement in my professional interactions between the United States and the metric-using countries. I often have to visualize the conversion from one system to the other depending on the current context of my professional engagements. This can be a disadvantage in critical high-pressure and time-sensitive engagements requiring a quick response. This could become further exacerbated for medical practitioners in response to world-centric responses to the COVID-19 pandemic.

As a good illustration of dual usage of measuring systems, a widespread news report in late September 1999 reported how the National Aeronautics and Space Administration (NASA) lost a $125 million Mars orbiter in a crash onto the surface of Mars because a Lockheed Martin engineering team used the English units of measurement, whereas the agency's team, based in Europe, used the more conventional metric system for a key operation of the spacecraft. The unit's mismatch prevented navigation information from transferring between the Mars Climate Orbiter spacecraft team at Lockheed Martin in Denver and the flight team at NASA's Jet Propulsion Laboratory in Pasadena, California. So, even at such a high-stakes scientific endeavor, nonstandardization of measuring units can create havoc.

The standardized structure and decimal features of the metric system made it well suited for scientific and engineering work. Consequently, it is not surprising that the rapid spread of the system coincided with an age of rapid technological development. In the United States, by the Act of Congress in 1866, it became "lawful throughout the United States of America to employ the weights and measures of the metric system in all contracts, dealings or court proceedings." However, the United States has remained a hold-out with respect to a widespread adoption of the metric system. Today, in some localities of the United States, both English and metric systems are used side by side.

In 1875, an international agreement, known as the Meter Convention, set up well-defined metric standards for length and mass and established permanent mechanisms to recommend and adopt further refinements in the metric system. This agreement, commonly called the "Treaty of the Meter" in the United States, was signed by 17 countries, including the United States. As a result of the Treaty, metric standards were constructed and distributed to each nation that ratified the Convention. Since 1893, the internationally adopted metric standards have served as the fundamental measurement standards of the United States, at least in theory, if not in practice.

By 1900 a total of 35 nations, including the major nations of continental Europe and most of South America, had officially accepted the metric system. In 1960, the General Conference on Weights and Measures, the diplomatic organization made up of the signatory nations to the Meter Convention, adopted an extensive revision and

# Essentials of Data Analytics

simplification of the system. The following seven units were adopted as the base units for the metric system:

1. Meter (for length)
2. Kilogram (for mass)
3. Second (for time)
4. Ampere (for electric current)
5. Kelvin (for thermodynamic temperature)
6. Mole (for amount of substance)
7. Candela (for luminous intensity)

Based on the general standardization described above, the name Système International d'Unités (International System of Units), with the international abbreviation SI, was adopted for the modern metric system. Throughout the world, measurement science research and development continue to develop more precise and easily reproducible ways of defining measurement units. The working organizations of the General Conference on Weights and Measures coordinate the exchange of information about the use and refinement of the metric system and make recommendations concerning improvements in the system and its related standards. Our daily lives are mostly ruled or governed by the measurements of length, weight, volume, and time.

## Fundamental Scientific Equations

Several fundamental equations govern how we do data analytics once we collect the data pertinent to a problem of interest. In modeling our experimental data, we often will need to develop our own best-fit equations. But there are cases where our modeling approach may be informed by standard scientific and engineering equations. Some of the seminal and fundamental theoretical scientific equations have emerged over the centuries. Perhaps, the most quoted and recognized in modern scientific literature is Einstein's equation.

## Einstein's Equation

$$E = mc^2$$

The fundamental relationship connecting energy, mass, and the speed of light emerges from Einstein's theory of special relativity, published in 1905. Showing the equivalence of mass and energy, it may be the most famous and beautiful equation in all of modern science. Its power was graphically demonstrated less than four decades later with the discovery of nuclear fission, a process in which a small amount of mass is converted to a very large amount of energy, precisely in accord with this equation.

## Einstein's Field Equation

$$R_{\mu\nu} - \frac{1}{2} g_{\mu\nu} R + \Lambda g_{\mu\nu} = 8\pi G T_{\mu\nu}$$

Einstein's elegant equation published in 1916 is the foundation for his theory of gravity, the theory of general relativity. The equation relates the geometrical curvature of space-time to the energy density of matter. The theory constructs an entirely new picture of space and time, out of which gravity emerges in the form of geometry and from which Newton's theory of gravity emerges as a limiting case. Einstein's field equation explains many features of modern cosmology, including the expansion of the universe and the bending of star light by matter, and it predicts black holes and gravitational waves. He introduced a cosmological constant in the equation, which he called his greatest blunder, but that quantity may be needed if, as recent observations suggest, the expansion of the universe is accelerating. A remaining challenge for physicists in the 21st century is to produce a fundamental theory uniting gravitation and quantum mechanics.

## Heisenberg's Uncertainty Principle

$$\Delta x \Delta p \geq \frac{h}{2}$$

Werner Heisenberg's matrix formulation of quantum mechanics led him to discover in 1927 that an irreducible uncertainty exists when simultaneously measuring the position and momentum of an object. Unlike classical mechanics, quantum mechanics requires that the more accurately the position of an object is known, the less accurately its momentum is known, and vice versa. The magnitude of that irreducible uncertainty is proportional to Planck's constant.

## Schrödinger Equation

$$i\hbar \frac{\partial \Psi}{\partial t} = H\Psi$$

In 1926, Erwin Schrödinger derived his nonrelativistic wave equation for the quantum mechanical motion of particles such as electrons in atoms. The probability density of finding a particle at a particular position in space is the square of the absolute value of the complex wave function, which is calculated from Schrödinger's equation. This equation accurately predicts the allowed energy levels for the electron in the hydrogen atom. With the use of modern computers, generalizations of this equation predict the properties of larger molecules and the behavior of electrons in complex materials.

## Dirac Equation

$$i\hbar \frac{\partial \Psi}{\partial t} = \left[ c\vec{\alpha}\left(\vec{p} - \vec{A}\right) + \beta mc^2 + e\Phi \right] \Psi$$

In 1928, Paul Dirac derived a relativistic generalization of Schrödinger's wave equation for the quantum mechanical motion of a charged particle in an electromagnetic field. His marvelous equation predicts the magnetic moment of the electron and the existence of antimatter.

# Essentials of Data Analytics

## Maxwell's Equations

$$\vec{\nabla} \cdot \vec{D} = p$$

$$\vec{\nabla} \times \vec{H} = \vec{J} + \frac{\partial \vec{D}}{\partial t}$$

$$\vec{\nabla} \times \vec{E} + \frac{\partial \vec{B}}{\partial t} = 0$$

$$\vec{\nabla} \cdot \vec{B} = 0$$

The fundamental equations explaining classical electromagnetism were developed over many years by James Clerk Maxwell and finished in his famous treatise published in 1873. His classical field theory provides an elegant framework for understanding electricity, magnetism, and propagation of light. Maxwell's theory was a major achievement of 19th-century physics, and it contained one of the clues that were used years later by Einstein to develop special relativity. Classical field theory was also the springboard for the development of quantum filed theory.

## Boltzmann's Equation for Entropy

$$S = k \ln W$$

Ludwig Boltzmann, one of the founders of statistical mechanics in the late 19th century, proposed that the probability for any physical state of macroscopic system is proportional to the number of ways in which the internal state of that system can be rearranged without changing the system's external properties. When more arrangements are possible, the system is more disordered. Boltzmann showed that the logarithm of the multiplicity of states of a system, or its disorder, is proportional to its entropy, and the constant of proportionality is Boltzmann's constant $k$. The second law of thermodynamics states that the total entropy of a system and its surroundings always increases as time elapses. Boltzmann's equation for entropy is carved on his grave.

## Planck–Einstein Equation

$$E = hv$$

The simple relation between the energy of a light quantum and the frequency of the associated light wave first emerged in a formula discovered in 1900 by Max Planck. He was examining the intensity of electromagnetic radiation emitted by the atoms in the walls of an enclosed cavity (a blackbody) at fixed temperature. He found that he could fit the experimental data by assuming that the energy associated with each mode of the electromagnetic field is an integral multiple of some minimum energy that is proportional to the frequency. The constant of proportionality, $h$, is known as Planck's constant. It is one of the most important fundamental numbers in physics.

In 1905, Albert Einstein recognized that Planck's equation implies that light is absorbed or emitted in discrete quanta, explaining the photoelectric effect and igniting the quantum mechanical revolution.

**Planck's Blackbody Radiation Formula**

$$u = \frac{8\pi h}{c^3} v^3 \left[ e^{\frac{h v}{kT}} - 1 \right]^{-1}$$

In studying the energy density of radiation in a cavity, Max Planck compared two approximate formulas: one for low frequency and one for high frequency. In 1900, using an ingenious extrapolation, he found his equation for the energy density of blackbody radiation, which reproduced experimental results. Seeking to understand the significance of his formula, he discovered the relation between energy and frequency known as Planck–Einstein equation.

**Hawking Equation for Black Hole Temperature**

$$T_{BH} = \frac{hc^3}{8\pi GMk}$$

Using insights from thermodynamics, relativist quantum mechanics, and Einstein's gravitational theory, Stephen Hawking predicted in 1974 the surprising result that gravitational black holes, which are predicted by general relativity, would radiate energy. His formula for the temperature of the radiating black hole depends on the gravitational constant, Planck's constant, the speed of light, and Boltzmann's constant. While Hawking radiation remains to be observed, his formula provides a tempting glimpse of the insights that will be uncovered in a unified theory combining quantum mechanics and gravity.

**Navier–Stokes Equation for a Fluid**

$$\rho \frac{\partial \vec{v}}{\partial t} + \rho (\vec{v} \cdot \vec{\nabla}) \vec{v} = -\vec{\nabla} p + \mu \nabla^2 \vec{v} + (\lambda + \mu) \vec{\nabla}(\vec{\nabla} \cdot \vec{v}) + \rho \vec{g}$$

The Navier–Stokes equation was derived in the 19th century from Newtonian mechanics to model viscous fluid flow. Its nonlinear properties make it extremely difficult to solve, even with the modern analytic and computational technique. However, its solutions describe a rich variety of phenomena including turbulence.

**Lagrangian for Quantum Chromodynamics**

$$L_{QDC} = -\frac{1}{4} F_a^{\mu\nu} \cdot F_{a\mu\nu} + \sum_f \overline{\Psi}_f \left[ i \slashed{\nabla} - g \slashed{A}_a t_a - m_f \right] \Psi_f$$

Relativistic quantum field theory had its first great success with quantum electrodynamics, which explains the interaction of charged particles with the quantized

# Essentials of Data Analytics

electromagnetic field. Exploration of non-Abelian gauge theories led next to the spectacular unification of the electromagnetic and weak interactions. Then, with insights developed from the quark model, quantum chromodynamics was developed to explain the strong interactions. This theory predicts that quarks are bound more tightly together as their separation increases, which explains why individual quarks are not seen directly in experiments. The standard model, which incorporates strong, weak, and electromagnetic interactions in a single quantum field theory, describes the interaction of quarks, gluons, and leptons, and has achieved remarkable success in predicting experimental results in elementary particle physics.

## Bardeen–Cooper–Schrieffer Equation for Superconductivity

$$T_c = 1.13\Theta e^{-\frac{1}{N(0)V}}$$

Superconductors are materials that exhibit no electrical resistance at low temperatures. In 1957 John Bardeen, Leon N. Cooper, and J. Robert Schrieffer applied quantum field theory with an approximate effective potential to explain this unique behavior of electrons in a superconductor. The electrons were paired and move collectively without resistance in the crystal lattice of the super-conducting material. The Bardeen–Cooper–Schrieffer (BCS) theory and its later generalizations predict a wide variety of phenomena that agree with experimental observations and have many practical applications. John Bardeen's contributions to solid-state physics also include inventing the transistor, made from semiconductors, with Walter Brattain and William Shockley in 1947.

## Josephson Effect

$$\frac{d(\Delta\varphi)}{dt} = \frac{2eV}{h}$$

In 1962, Brian Josephson made the remarkable prediction that electric current could flow between two thin pieces of superconducting material separated by a thin piece of insulating material without application of a voltage. Using the BCS theory of superconductivity, he also predicted that if a voltage difference were maintained across the junction, there would be an alternating current with a frequency related to the voltage and Planck's constant. The presence of magnetic fields influences the Josephson effect, allowing it to be used to measure very weak magnetic fields approaching the microscopic limit set by quantum mechanics.

## Fermat's Last Theorem

$$x^n + y^n = z^n$$

While studying the properties of whole numbers, or integers, the French mathematician Pierre de Fermat wrote in 1637 that it is impossible for the cube of an integer to be written as the sum of the cubes of two other integers. More generally, he stated that it is impossible to find such a relation between three integers for any integral power greater than two. He went on to write a tantalizing statement in the margin of his copy of a Latin translation of Diophantus's *Arithemetica:* "I have a truly marvelous demonstration of this proposition, which this margin is too narrow to contain." It took over 350 years to prove Fermat's simple conjecture. The feat was achieved by Andrew Wiles in 1994 with a "tour de force" proof of many pages using newly developed techniques in number theory. It is noteworthy that many researchers, mathematicians, and scholars toiled for almost four centuries before a credible proof of Fermat's last theorem was found. Indeed, the lead editor of this handbook, as a Mathematics graduate student in the early 1980s, was introduced to the problem during his Advanced Calculus studies under Professor Reginald Mazeres at Tennessee Technological University in 1980. Like many naïve researchers before him, he struggled with the problem as a potential thesis topic for six months before abandoning it to pursue a more doable topic in predictive time series modeling.

## Methods for Data Measurement and Comparison

There are two basic methods of measurement:

1. *Direct comparison* with either a primary or a secondary standard
2. *Indirect comparison* with a standard through the use of a calibrated system

### Direct Comparison

How do you measure the length of a cold-rolled bar? You probably use a steel tape. You compare the bar's length with a standard. The bar is so many feet long because that many units on your standard have the same length as the bar. You have determined this by making a direct comparison. Although you do not have access to the primary standard defining the unit, you manage very well with a secondary standard. Primary measurement standards have the least amount of uncertainty compared to the certified value and are traceable directly to the SI. Secondary standards, on the other hand, are derived by assigning value by comparison to a primary standard.

In some respect, measurement by direct comparison is quite common. Many length measurements are made in this way. In addition, time of day is usually determined by comparison, with a watch used as a secondary standard. The watch goes through its double cycle, in synchronization with the Earth's rotation. Although, in this case, the primary standard is available to everyone, the watch is more convenient because its works on cloudy days, indoors, outdoors, in daylight, and in the dark (at night). It is also more precise. That is, its resolution is better. In addition, if well regulated, the watch is more accurate, because the Earth does not rotate at a uniform speed. It is seen, therefore, that in some cases, a secondary standard is actually more useful than the primary standard.

Measuring by direct comparison implies stripping the measurement problem to its barest essentials. However, the method is not always the most accurate or the best. The human senses are not equipped to make direct comparisons of all quantities with equal facility. In many cases, they are not sensitive enough. We can make direct length comparisons using a steel rule with a level of precision of about 0.01 inch. Often, we wish for a greater accuracy, in which case we must call for additional assistance from some calibrated measuring system.

**Indirect Comparison**

While we can do a reasonable job through direct comparison of length, how well can we compare masses, for example? Our senses enable us to make rough comparisons. We can lift a pound of meat and compare its effect with that of some unknown mass. If the unknown is about the same weight, we may be able to say that it is slightly heavier, or perhaps, not quite as heavy as our "standard" pound, but we could never be certain that the two masses were the same, even say within one ounce. Our ability to make this comparison is not as good as it is for the displacement of the mass. Our effectiveness in coming close to the standard is related to our ability to "gage" the relative impacts of mass on our ability to displace the mass. This brings to mind the common riddles of "Which weighs more? A pound of feathers or a pound of stones?" Of course, both weigh the same with respect to the standard weight of "pound."

In making most engineering measurements, we require the assistance of some form of measuring system, and measurement by direct comparison is less general than measurement by indirect comparison.

A generic measurement sequence can involve the following steps:

Identify the variable to measure or for which data is to be collected.
Take the actual measurement.
Analyze the measurement (or data).
Interpret the measurement in the context of the prevailing practical application.
Generate information from the data interpretation.
Make actionable decision from the information.
Implement the action.
Communicate the action.
Do an assessment of the outcome for feedback purpose.
Institute process improvement for the next cycle of data engagement.

## Data Measurement Scales

Every decision requires data collection, measurement, and analysis. In practice, we encounter different types of measurement scales depending on the particular items of interest. Data may need to be collected on decision factors, costs, performance levels, outputs, and so on. The different types of data measurement scales that are applicable are presented below.

### Nominal Scale of Measurement

Nominal scale is the lowest level of measurement scales. It classifies items into categories. The categories are mutually exclusive and collectively exhaustive. That is, the categories do not overlap, and they cover all possible categories of the characteristics being observed. For example, in the analysis of the critical path in a project network, each job is classified as either critical or not critical. Gender, type of industry, job classification, and color are examples of measurements on a nominal scale.

### Ordinal Scale of Measurement

Ordinal scale is distinguished from a nominal scale by the property of order among the categories. An example is the process of prioritizing project tasks for resource allocation. We know that first is above second, but we do not know how far above. Similarly, we know that better is preferred to good, but we do not know by how much. In quality control, the ABC classification of items based on the Pareto distribution is an example of a measurement on an ordinal scale.

### Interval Scale of Measurement

Interval scale is distinguished from an ordinal scale by having equal intervals between the units of measurement. The assignment of priority ratings to project objectives on a scale of 0–10 is an example of a measurement on an interval scale. Even though an objective may have a priority rating of zero, it does not mean that the objective has absolutely no significance to the project team. Similarly, the scoring of zero on an examination does not imply that a student knows absolutely nothing about the materials covered by the examination. Temperature is a good example of an item that is measured on an interval scale. Even though there is a zero point on the temperature scale, it is an arbitrary relative measure. Other examples of interval scale are IQ measurements and aptitude ratings.

### Ratio Scale Measurement

Ratio scale has the same properties of an interval scale but with a true zero point. For example, an estimate of zero-time unit for the duration of a task is a ratio scale measurement. Other examples of items measured on a ratio scale are cost, time, volume, length, height, weight, and inventory level. Many of the items measured in engineering systems will be on a ratio scale.

Another important aspect of measurement involves the classification scheme used. Most systems will have both quantitative and qualitative data. Quantitative data require that we describe the characteristics of the items being studied numerically. Qualitative data, on the other hand, are associated with attributes that are not measured numerically. Most items measured on the nominal and ordinal scales will normally be classified into the qualitative data category, whereas those measured on the interval and ratio scales will normally be classified into the quantitative data category. The implication for engineering system control is that qualitative data can lead to bias in the control mechanism because qualitative data are subject to the personal views and interpretations of the person using the data. As much as possible, data for an engineering systems control should be based on a quantitative measurement.

# Essentials of Data Analytics

As a summary, examples of the four types of data measurement scales are presented below:

- Nominal scale (attribute of classification): color, gender, design type
- Ordinal scale (attribute of order): first, second, low, high, good, better
- Interval scale (attribute of relative measure): intelligence quotient, grade point average, temperature
- Ratio (attribute of true zero): cost, voltage, income, budget

Notice that temperature is included in the "relative" category rather the "true zero" category. Even though there are zero temperature points on the common temperature scales (i.e., Fahrenheit, Celsius, and Kelvin), those points are experimentally or theoretically established. They are not true points as one might find in a counting system.

## REFERENCE UNITS OF MEASUREMENTS

Some common units of measurement for reference purposes are provided below:

**Acre.** An area of 43,560 square feet.
**Agate.** 1/14 inch (used in printing for measuring column length).
**Ampere.** Unit of electric current.
**Astronomical (A.U.).** 93,000,000 miles; the average distance of the Earth from the sun (used in astronomy).
**Bale.** A large bundle of goods. In the United States, approximate weight of a bale of cotton is 500 pounds. The weight of a bale may vary from country to country.
**Board foot.** 144 cubic inches (12 by 12 by 1 used for lumber).
**Bolt.** 40 yards (used for measuring cloth).
**Btu.** British thermal unit; the amount of heat needed to increase the temperature of one pound of water by 1°F (252 calories).
**Carat.** 200 mg or 3,086 troy; used for weighing precious stones (originally the weight of a seed of the carob tree in the Mediterranean region). See also *Karat*.
**Chain.** 66 feet; used in surveying (1 mile = 80 chains).
**Cubit.** 18 inches (derived from the distance between the elbow and the tip of the middle finger).
**Decibel**. Unit of relative loudness.
**Freight Ton.** 40 cubic feet of merchandise (used for cargo freight).
**Gross.** 12 dozens (144).
**Hertz.** Unit of measurement of electromagnetic wave frequencies (measures cycles per second).
**Hogshead.** Two liquid barrels or 14,653 cubic inches.
**Horsepower.** The power needed to lift 33,000 pounds a distance of one foot in one minute (about 1½ times the power an average horse can exert); used for measuring the power of mechanical engines.

**Karat.** A measure of the purity of gold. It indicates how many parts out of 24 are pure. 18 karat gold is ¾ pure gold.
**Knot.** The rate of speed of 1 nautical mile per hour; used for measuring the speed of ships (not distance).
**League.** Approximately 3 miles.
**Light-year.** 5,880,000,000,000 miles; the distance traveled by light in one year at the rate of 186,281.7 miles per second; used for measuring the interstellar space.
**Magnum.** Two-quart bottle; used for measuring wine.
**Ohm.** Unit of electrical resistance.
**Parsec.** Approximately 3.26 light-years of 19.2 trillion miles; used for measuring interstellar distances.
**Pi (π).** 3.14159265+; the ratio of the circumference of a circle to its diameter.
**Pica.** 1/6 inch or 12 points; used in printing for measuring the column width.
**Pipe.** 2 hogsheads; used for measuring wine and other liquids.
**Point.** 0.013837 (approximately 1/72 inch or 1/12 pica); used in printing for measuring type size.
**Quintal.** 100,000 g or 220.46 pounds avoirdupois.
**Quire.** 24 or 25 sheets; used for measuring paper (20 quires is one ream).
**Ream.** 480 or 500 sheets; used for measuring paper.
**Roentgen.** Dosage unit of radiation exposure produced by X-rays.
**Score.** 20 units.
**Span.** 9 inches or 22.86 cm; derived from the distance between the end of the thumb and the end of the little finger when both are outstretched.
**Square.** 100 square feet; used in building.
**Stone.** 14 pounds avoirdupois in Great Britain.
**Therm.** 100,000 Btu's.
**Township.** U.S. land measurement of almost 36 square miles; used in surveying.
**Tun.** 252 gallons (sometimes larger); used for measuring wine and other liquids.
**Watt.** Unit of power.

## COMMON CONSTANTS

Speed of light: $2.997,925 \times 10^{10}$ cm/sec ($983.6 \; 10^6$ ft/sec; 186,284 miles/sec)
Velocity of sound: 340.3 m/sec (1,116 ft/sec)
Gravity (acceleration): 9.80665 m/sec$^2$ (32.174 ft/sec$^2$; 386.089 inches/sec$^2$)

## NUMERIC DATA REPRESENTATION

Exponentiation is essential for data measurement and presentation for both small and large numbers. For the purpose of data analytics, the data exponentiation system is important for analysts. The standard exponentiation numbers and prefixes are presented below:

# Essentials of Data Analytics

| | |
|---|---|
| yotta ($10^{24}$) | 1 000 000 000 000 000 000 000 000 |
| zetta ($10^{21}$) | 1 000 000 000 000 000 000 000 |
| exa ($10^{18}$) | 1 000 000 000 000 000 000 |
| peta ($10^{15}$) | 1 000 000 000 000 000 |
| tera ($10^{12}$) | 1 000 000 000 000 |
| giga ($10^{9}$) | 1 000 000 000 |
| mega ($10^{6}$) | 1 000 000 |
| kilo ($10^{3}$) | 1 000 |
| hecto ($10^{2}$) | 100 |
| deca ($10^{1}$) | 10 |
| deci ($10^{-1}$) | 0.1 |
| centi ($10^{-2}$) | 0.01 |
| milli ($10^{-3}$) | 0.001 |
| micro ($10^{-6}$) | 0.000 001 |
| nano ($10^{-9}$) | 0.000 000 001 |
| pico ($10^{-12}$) | 0.000 000 000 001 |
| femto ($10^{-15}$) | 0.000 000 000 000 001 |
| atto ($10^{-18}$) | 0.000 000 000 000 000 001 |
| zepto ($10^{-21}$) | 0.000 000 000 000 000 000 001 |
| yocto ($10^{-24}$) | 0.000 000 000 000 000 000 000 001 |

## THE LANGUAGE OF DATA ANALYTICS

Mathematicians and statisticians are often accused of speaking in their own strange "Greek" language when presenting their work. To do and appreciate data analytics, we must learn to recognize and use the common notations and symbols commonly used in mathematics and statistics. For this reason, we are presenting this reference material up front here rather than in an appendix. Table 1.1 presents a tabulation of Greek symbols often used for data analytics, whereas Table 1.2 presents the common Roman numerals. This entire book is billed as reference handbook for equations and formulas for data analytics. As such, the reference equations and formulas are presented as the inherent body of the book.

## QUICK REFERENCE FOR MATHEMATICAL EQUATIONS

Presented below is a collection of quick references for mathematical equations and formulas:

$$\sum_{n=0}^{\infty} \frac{x^n}{n!} = e^x$$

$$\sum_{n=0}^{\infty} \frac{x^n}{n} = \ln\left(\frac{1}{1-x}\right)$$

## TABLE 1.1
### Greek Symbols for Data Analytics

| Capital | Lowercase | Greek Name | Pronunciation | English |
|---|---|---|---|---|
| A | $\alpha$ | Alpha | al-fah | a |
| B | $\beta$ | Beta | bay-tah | b |
| Γ | $\gamma$ | Gamma | gam-ah | g |
| Δ | $\delta$ | Delta | del-tah | d |
| E | $\varepsilon$ | Epsilon | ep-si-lon | e |
| Z | $\zeta$ | Zeta | zat-tah | z |
| H | $\eta$ | Eta | ay-tah | h |
| Θ | $\theta$ | Theta | thay-tah | th |
| I | $\iota$ | Iota | eye-oh-tah | i |
| K | $\kappa$ | Kappa | cap-ah | k |
| Λ | $\lambda$ | Lambda | lamb-da | l |
| M | $\mu$ | Mu | mew | m |
| N | $\nu$ | Nu | new | n |
| Ξ | $\xi$ | Xi | sah-eye | x |
| O | $o$ | Omicron | oh-mi-cron | o |
| Π | $\pi$ | Pi | pie | p |
| P | $\rho$ | Rho | roe | r |
| Σ | $\sigma$ | Sigma | sig-mah | s |
| T | $\tau$ | Tau | tah-hoe | t |
| Υ | $\upsilon$ | Upsilon | oop-si-lon | u |
| Φ | $\varphi$ | Phi | fah-eye | ph |
| X | $\chi$ | Chi | kigh | ch |
| Ψ | $\psi$ | Psi | sigh | ps |
| Ω | $\Omega$ | Omega | Oh-mega | o |

## TABLE 1.2
### Roman Numerals for Data Analytics

| | | | | | | | | |
|---|---|---|---|---|---|---|---|---|
| 1 | I | 14 | XIV | 27 | XXVII | 150 | CL |
| 2 | II | 15 | XV | 28 | XXVIII | 200 | CC |
| 3 | III | 16 | XVI | 29 | XXIX | 300 | CCC |
| 4 | IV | 17 | XVII | 30 | XXX | 400 | CD |
| 5 | V | 18 | XVIII | 31 | XXXI | 500 | D |
| 6 | VI | 19 | XIX | 40 | XL | 600 | DC |
| 7 | VII | 20 | XX | 50 | L | 700 | DCC |
| 8 | VIII | 21 | XXI | 60 | LX | 800 | DCCC |
| 9 | IX | 22 | XXII | 70 | LXX | 900 | CM |
| 10 | X | 23 | XXIII | 80 | LXXX | 1000 | M |
| 11 | XI | 24 | XXIV | 90 | XC | 1600 | MDC |
| 12 | XII | 25 | XXV | 100 | C | 1700 | MDCC |
| 13 | XIII | 26 | XXVI | 101 | CI | 1900 | MCM |

# Essentials of Data Analytics

$$\sum_{n=0}^{k} x^n = \frac{x^{k+1}-1}{x-1}, \quad x \neq 1$$

$$\sum_{n=1}^{k} x^n = \frac{x-x^{k+1}}{1-x}, \quad x \neq 1$$

$$\sum_{n=2}^{k} x^n = \frac{x^2-x^{k+1}}{1-x}, \quad x \neq 1$$

$$\sum_{n=0}^{\infty} p^n = \frac{1}{1-p}, \quad \text{if } |p|<1$$

$$\sum_{n=0}^{\infty} nx^n = \frac{x}{(1-x)^2}, \quad x \neq 1$$

$$\sum_{n=0}^{\infty} n^2 x^n = \frac{2x^2}{(1-x)^3} + \frac{x}{(1-x)^2} = \frac{x(1+x)}{(1-x)^3}, \quad |x|<1$$

$$\sum_{n=0}^{\infty} n^3 x^n = \frac{6x^3}{(1-x)^4} + \frac{6x^2}{(1-x)^3} + \frac{x}{(1-x)^2}, \quad |x|<1$$

$$\sum_{n=0}^{M} nx^n = \frac{x\left[1-(M+1)x^M + Mx^{M+1}\right]}{(1-x)^2}, \quad |x|<1$$

$$\sum_{x=0}^{\infty} \binom{r+x-1}{x} u^x = (1-u)^{-r}, \quad \text{if } |u|<1$$

$$\sum_{k=1}^{\infty} (-1)^{k+1} \frac{1}{k} = 1 - \frac{1}{2} + \frac{1}{3} - \frac{1}{4} + \frac{1}{5} - \frac{1}{6} + \cdots = \ln 2$$

$$\sum_{k=1}^{\infty} (-1)^{k+1} \frac{1}{(2k-1)} = 1 - \frac{1}{3} + \frac{1}{5} - \frac{1}{7} + \frac{1}{9} - \cdots = \frac{\pi}{4}$$

$$\sum_{k=0}^{\infty} (-1)^k x^k = \frac{1}{1+x}, \quad -1<x<1$$

$$\sum_{k=1}^{n}(-1)^k\binom{n}{k}=1, \quad \text{for } n\geq 2$$

$$\sum_{k=0}^{n}\binom{n}{k}^2=\binom{2n}{n}$$

$$\sum_{k=1}^{n}k=1+2+3+\cdots+n=\frac{n(n+1)}{2}$$

$$\sum_{k=1}^{n}k^2=1+4+9+\cdots+n^2=\frac{n(n+1)(2n+1)}{6}$$

$$\sum_{k=0}^{n-1}k^2x^k=\frac{(x-1)^2 n^2 x^n - 2(x-1)nx^{n+1}+x^{n+2}-x^2+x^{n+1}-x}{(x-1)^3}$$

$$\sum_{k=1}^{n}k^3=1+8+27+\cdots+n^3=\left(\frac{n(n+1)}{2}\right)^2$$

$$\sum_{k=1}^{n}(2k)=2+4+6+\cdots+2n=n(n-1)$$

$$\sum_{k=1}^{n}(2k-1)=1+3+5+\cdots+(2n-1)=n^2$$

$$\sum_{k=0}^{\infty}(a+kd)r^k=a+(a+d)r+(a+2d)r^2+\cdots=\frac{a}{1-r}+\frac{rd}{(1-r)^2}$$

$$\sum_{k=1}^{n}k^3=1+8+27+\cdots+n^3=\frac{n^2(n+1)^2}{4}=\left[\frac{n(n+1)}{2}\right]^2=\left[\sum_{k=1}^{n}k\right]^2$$

$$\sum_{x=1}^{\infty}\frac{1}{x}=1+\frac{1}{2}+\frac{1}{3}+\cdots(\text{does not converge})$$

$$\sum_{m=0}^{k}ma^m=\frac{a}{(1-a)^2}\left[1-(k+1)a^k+ka^{k+1}\right]=\sum_{m=1}^{k}ma^m$$

$$\sum_{k=0}^{n}(1)=n$$

$$\sum_{k=0}^{n}\binom{n}{k}=2^{n}$$

$$(a+b)^{n}=\sum_{k=0}^{n}\binom{n}{k}a^{k}b^{n-k}$$

$$\prod_{n=1}^{\infty}a_{n}=e^{\left(\sum_{n=1}^{\infty}\ln(a_{n})\right)}$$

$$\ln\left(\prod_{n=1}^{\infty}a_{n}\right)=\sum_{n=1}^{\infty}\ln a_{n}$$

$$\ln(x)=\sum_{k=1}^{\infty}\frac{1}{k}\left(\frac{x-1}{x}\right)^{k}, \quad x\geq\frac{1}{2}$$

$$\lim_{h\to\infty}(1+h)^{1/h}=e$$

$$\lim_{n\to\infty}\left(1+\frac{x}{n}\right)^{n}=e^{-x}$$

$$\lim_{n\to\infty}\sum_{k=0}^{n}\frac{e^{-n}n^{r}}{K!}=\frac{1}{2}$$

$$\lim_{k\to\infty}\left(\frac{x^{k}}{k!}\right)=0$$

$$|x+y|\leq|x|+|y|$$

$$|x-y|\geq|x|-|y|$$

$$\ln(1+x)=\sum_{k=1}^{\infty}(-1)^{k+1}\left(\frac{x^{k}}{k}\right), \quad \text{if } -1<x\leq 1$$

$$\Gamma\left(\frac{1}{2}\right)=\sqrt{\pi}$$

$$\Gamma(\alpha+1)=\alpha\Gamma(\alpha)$$

$$\Gamma\left(\frac{n}{2}\right) = \frac{\sqrt{\pi}(n-1)!}{2^{n-1}\left(\frac{n-1}{2}\right)!}, \quad n \text{ odd}$$

$$\Gamma(n) = \int_0^\infty e^{-x} x^{n-1} \, dx$$

$$\binom{n}{2} = \frac{1}{2}(n^2 - n) = \sum_{k=1}^{n-1} k$$

$$\binom{n+1}{2} = \binom{n}{2} + n$$

$$2.4.6.8\ldots 2n = \prod_{k=1}^{n} 2k = 2^n n!$$

$$1.3.5.7\ldots(2n-1) = \frac{(2n-1)!}{2^{2n-2}(2n-2)!} = \frac{2n-1}{2^{2n-2}}$$

Derivation of closed form expression for $\sum_{k=1}^{n} kx^k$

$$\sum_{k=1}^{n} kx^k = x \sum_{k=1}^{n} kx^{k-1}$$

$$= x \sum_{k=1}^{n} \frac{d}{dx}\left[x^k\right]$$

$$= x \frac{d}{dx}\left[\sum_{k=1}^{n} x^k\right]$$

$$= x \frac{d}{dx}\left[\frac{x(1-x^n)}{1-x}\right]$$

$$= x \left[\frac{(1-(n+1)x^n)(1-x) - x(1-x^n)(-1)}{(1-x)^2}\right]$$

$$= \frac{x\left[1-(n+1)x^n + nx^{n+1}\right]}{(1-x)^2}, \quad x \neq 1$$

# Essentials of Data Analytics

## Derivation of the Quadratic Formula

Formula:

$$ax^2 + bx + c = 0$$

Solution:

$$x = \frac{-b \pm \sqrt{b^2 - 4ac}}{2a}$$

If $b^2 - 4ac < 0$, the roots are complex.
If $b^2 - 4ac > 0$, the roots are real.
If $b^2 - 4ac = 0$, the roots are real and repeated.

Formula:

$$ax^2 + bx + c = 0$$

Dividing both sides by "$a$," ($a \neq 0$)

$$x^2 + \frac{b}{a}x + \frac{c}{a} = 0$$

Note if $a = 0$, the solution to $ax^2 + bx + c = 0$ is $x = -\frac{c}{b}$.
Rewrite

$$x^2 + \frac{b}{a}x + \frac{c}{a} = 0$$

as

$$\left(x + \frac{b}{2a}\right)^2 - \frac{b^2}{4a^2} + \frac{c}{a} = 0$$

$$\left(x + \frac{b}{2a}\right)^2 = \frac{b^2}{4a^2} - \frac{c}{a} = \frac{b^2 - 4ac}{4a^2}$$

$$x + \frac{b}{2a} = \pm\sqrt{\frac{b^2 - 4ac}{4a^2}} = \pm\frac{\sqrt{b^2 - 4ac}}{2a}$$

$$x = -\frac{b}{2a} \pm \sqrt{\frac{b^2 - 4ac}{2a}}$$

$$x = \frac{-b \pm \sqrt{b^2 - 4ac}}{2a}$$

# REFERENCE

Badiru, Adedeji B. (1981), "Stochastic Model of Energy Consumption at Tennessee Technological University," Masters Thesis, Cookeville, Tennessee.

# 2 Empirical Model Building

Believe me or not, the model tells you the facts.

## INTRODUCTION TO THE MODEL ENVIRONMENT

This chapter presents concept frameworks for building a model based on the available data. It addresses data regression and data projection. It also addresses the mathematical steps to building an empirical model. Models are not crystal balls. Yet, they are relied upon to get glimpse of the future, even if it is a fuzzy view. Models are forecasts. In this regard, they are better than nothing. Therein lies the efficacy of data analytics. Figure 2.1 illustrates the empirical modeling environment representing data linkages of the past, present, and future projections. The simplest and most common types of models are those representing growth and decay. An example is as follows:

$$X_{(j+1)} = (1+\alpha)X_j$$

This is a case of the next element at time $j+1$ being derived from the preceding element at time $j$ at the behest of a multiplier factor of $(1+\alpha)$. With an appropriate choice of the value of $\alpha$, this simple equation can be made to generate a growth or a decay. Another simple example is the exponential form presented below:

$$y = e^t$$

COVID-19 presents a particularly challenging environment for modeling. One difficulty is that we don't yet have enough data to generate reliable or meaningful mathematical models. Another difficulty is that, for now, COVID-19 is still mostly stochastic and nonlinear, leading to intractable patterns for a modeling exercise. However, as the pandemic moves on, stunted or uninhibited, we will acquire more data to make COVID-19 modeling more reliable. The premise of this book is to

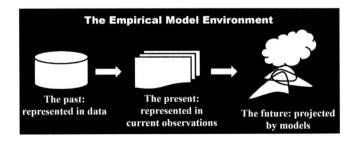

**FIGURE 2.1** Model environment representing data linkages of the past, present, and future.

provide a collection of the mathematical equations, formulas, and techniques now for the purpose of getting researchers and practitioners ready for that future state. As more incidents happen and are recorded, we will achieve more data agility to make modeling possible and more reliable.

## STATE-SPACE MODELING

The state space of a dynamical system is the set of all possible states of the system. Each coordinate of the state space is a state variable, and the values of all the state variables completely describe the state of the system. Classical control system focuses on control of the dynamics of mechanical objects, such as a pump, electrical motor, turbine, and rotating wheel. The mathematical basis for such control systems can be adapted (albeit in iconic formats) for general data analytics. This is because system transitions are characterized by inputs, variables, processing, control, feedback, and output. This is represented graphically by input–process–output relationship block diagrams. Mathematically, it can be represented as

$$z = f(x) + \varepsilon$$

where
$z$ is the output.
$f()$ is the functional relationship.
$\varepsilon$ is the error component (noise, disturbance, etc.).

For multivariable cases, the mathematical expression is represented as vector–matrix functions as shown in the following:

$$\mathbf{Z} = \mathbf{f}(X) + \mathbf{E}$$

where
each term is a matrix.

$\mathbf{Z}$ is the output vector.
$\mathbf{f}(\cdot)$ is the input vector.
$\mathbf{E}$ is the error vector.

Regardless of the level or form of mathematics used, all systems exhibit the same input–process–output characteristics, either quantitatively or qualitatively, as they transition from one state to another. System objectives are achieved by state-to-state transformations, where a subsequent state is derived from the characteristics of the preceding state. This simple representation can be expanded to cover several components within the realm of data analytics. Hierarchical linking of system elements provides an expanded transformation structure. The system state can be expanded in accordance with implicit requirements or explicit impositions. These requirements might include grouping of system elements, precedence linking (both technical and procedural), required communication links, and reporting requirements. The actions

# Empirical Model Building

to be taken at each state depend on the prevailing system conditions. The nature of subsequent alternate states depends on what actions are implemented. Sometimes, there are multiple paths that can lead to the end result. At other times, there exists only one unique path to the end result. In conventional practice, the characteristics of the future states can only be recognized after the fact, thus making it impossible to develop adaptive plans. If we describe a product by $P$ state variables $s_i$, then the composite state of the product at any given time can be represented by a vector $\mathbf{S}$ containing $P$ elements. That is,

$$\mathbf{S} = \{s_1, s_2, K, s_P\}$$

The components of the state vector could represent either quantitative or qualitative variables (e.g., cost, energy, color, time). We can visualize every state vector as a point in the $M$-dimensional state space. The representation is unique since every state vector corresponds to one and only one point in the state space.

Suppose we have a set of actions (transformation agents) that we can apply to a product information space so as to change it from one state to another within the system state space. The transformation will change a state vector into another state vector. For example, for a product development application, a transformation may be a change in raw material or a change in design approach. Suppose we let $T_k$ be the $k$th type of transformation. If $T_k$ is applied to the product when it is in state $\mathbf{S}$, the new state vector will be $T_k(\mathbf{S})$, which is another point in the state space. The number of transformations (or actions) available for a product may be finite or countably infinite. We can construct trajectories that describe the potential states of a product evolution as we apply successive transformations. Each transformation may be repeated as many times as needed. Given an initial state $\mathbf{S}_0$, the sequence of state vectors is represented by the following:

$$\mathbf{S}_1 = T_1(\mathbf{S}_0)$$

$$\mathbf{S}_2 = T_2(\mathbf{S}_1)$$

$$\mathbf{S}_3 = T_3(\mathbf{S}_2)$$

$$\ldots$$

$$\mathbf{S}_n = T_n(\mathbf{S}_{n-1})$$

The final state, $\mathbf{S}_n$, depends on the initial state $\mathbf{S}$ and the effects of the actions applied. Apart from the function of prediction, data analytics can also have the utility of helping to determine the decision paths for resource allocation. In the hypothetical framework in Figure 2.2, the resource base is distributed along priority paths based on predetermined weighting factors. The weighting factors are used to compute the relative number of resource units going into each resource budget. A resource allocation measure can be viewed as the action applied within a state space to move a system from one state to the next state. Thus, we can have distributive data analytics for resource allocation.

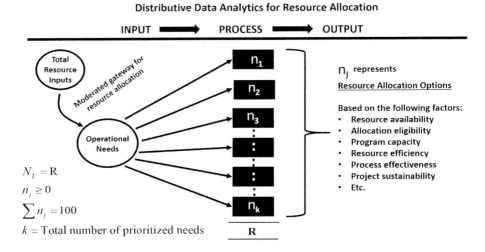

FIGURE 2.2  Graphical framework for distributive data analytics.

## CALCULUS REFERENCE FOR DATA ANALYTICS

Mathematics, simple or advanced, is the backbone of data analytics. Calculus, in particular, is essential for data analytics. A simple calculus reference here puts the basic concepts at the fingertip of the reader. Calculus is the mathematical study of continuous change, in the same way that geometry is the study of shape and algebra is the study of generalizations of arithmetic operations. Calculus presents the changes between values that are related by a function, such as what may be of interest in trend analysis in data analytics. Put simply, Calculus is the mathematics of rate of change, as in tracking functional changes in the rate of coronavirus pandemic infections. The rate of change computation is accomplished by the process of taking the "derivative" of a function or "differentiating" the function. For example, the rate of change of a constant (e.g., the number 5) is 0. The rate of change of the variable $x$ (actually $1x$) is 1. The definition of a derivative comes from taking the limit of the slope formula as the two points on a function get closer and closer together. Thus, the derivative is used to determine the slope of a curve at a particular point, which is an item of interest in straightening, bending, or reversing the curve of COVID-19 incidents. The converse of taking the derivative of a function or differentiating the function is finding the integral of the curve (i.e., differentiating the function). In other words, differentiation is the action of computing a derivative. The derivative of a function $y = f(x)$ of a variable $x$ is a measure of the rate at which the value $y$ of the function changes with respect to the change of the variable $x$. That rate of change is called the derivative of the function $f$ with respect to the variable $x$. All these basic concepts are essential in building data analytics models and performing functional analysis on the models for predictive, descriptive, or prescriptive purposes. Table 2.1 summarizes some basic integration guide that may be useful in systems modeling for data analytics.

# TABLE 2.1
## Derivatives and antiderivatives for selected functions

| Derivative | Integral (Antiderivative) |
|---|---|
| $\frac{d}{dx} n = 0$ | $\int 0\, dx = C$ |
| $\frac{d}{dx} x = 1$ | $\int 1\, dx = x + C$ |
| $\frac{d}{dx} e^x = e^x$ | $\int e^x = e^x + C$ |
| $\frac{d}{dx} \ln n = \frac{1}{x}$ | $\int \frac{1}{x} dx = \ln x + Cs$ |
| $\frac{d}{dx} n^x = n^x \ln n$ | $\int n^x dx = \frac{n^x}{\ln n} + C$ |
| $\frac{d}{dx} \sin x = \cos x$ | $\int \cos x\, dx = \sin x + C$ |
| $\frac{d}{dx} \cos x = -\sin x$ | $\int \sin x\, dx = -\cos x + C$ |
| $\frac{d}{dx} \tan x = \sec^2 x$ | $\int \sec^2 x\, dx = \tan x + C$ |
| $\frac{d}{dx} \cot x = -\csc^2 x$ | $\int \csc^2 x\, dx = -\cot x + C$ |
| $\frac{d}{dx} \sec x = \sec x \tan x$ | $\int \tan x \sec x\, dx = \sec x + C$ |
| $\frac{d}{dx} \csc x = -\csc x \cot x$ | $\int \cot x \csc x\, dx = -\csc x + C$ |
| $\frac{d}{dx} \arcsin x = -\frac{1}{\sqrt{1-x^2}}$ | $\int \frac{1}{\sqrt{1-x^2}} dx = \arcsin x + C$ |
| $\frac{d}{dx} \arccos x = -\frac{1}{\sqrt{1-x^2}}$ | $\int -\frac{1}{\sqrt{1-x^2}} dx = \arccos x + C$ |
| $\frac{d}{dx} \arctan x = \frac{1}{1+x^2}$ | $\int \frac{1}{1+x^2} dx = \arctan x + C$ |
| $\frac{d}{dx} \operatorname{arccot} x = -\frac{1}{1+x^2}$ | $\int -\frac{1}{1+x^2} dx = \operatorname{arccot} x + C$ |
| $\frac{d}{dx} \operatorname{arcsec} x = \frac{1}{x\sqrt{x^2-1}}$ | $\int \frac{1}{x\sqrt{x^2-1}} dx = \operatorname{arcsec} x + C$ |
| $\frac{d}{dx} \operatorname{arccsc} x = -\frac{1}{x\sqrt{x^2-1}}$ | $\int -\frac{1}{x\sqrt{x^2-1}} = \operatorname{arccsc} x + C$ |

## INTEGRATION RULES

The sum rule integrates long expressions term by term.

$$\int [f(x) + g(x)] dx = \int f(x)\, dx + \int g(x)\, dx$$

The constant multiple rule moves a constant outside of an integral before integration.

$$\int nf(x)\,dx = n\int f(x)\,dx$$

The power rule integrates any real power of $x$ (except $-1$).

$$\int x^n\,dx = \frac{x^{n-1}}{n+1} + C, \text{ where } n \neq 1$$

## SOLVING INTEGRALS WITH VARIABLE SUBSTITUTION

Step 1: Declare a variable $u$ and set it equal to an algebraic expression that appears in the integral, and then substitute $u$ for this expression in the integral.

Step 2: Differentiate $u$ to find $\frac{du}{dx}$ and then isolate all $x$ variables on one side of the equality sign.

Step 3: Make another substitution to change $dx$ and all other occurrences of $x$ in the integral to an expression that includes $du$.

Step 4: Integrate by using $u$ as the new variable of integration.

Step 5: Now, express the answer in terms of $x$.

## RIEMANN INTEGRAL

*Riemann integral* is the limit of the *Riemann* sums of a continuous function as the partitions get smaller and smaller. This approach is also applicable to functions that are not too seriously discontinuous. The Riemann sum formula provides a precise definition of the definite integral as the limit of an infinite series:

$$\int_a^b f(x)\,dx = \lim_{n\to\infty} \sum_{i=1}^n f(x_i)\left(\frac{b-a}{n}\right)$$

## INTEGRATION BY PARTS

To evaluate an integral by using integration by parts, follow these steps:

1. Decompose the entire integral (including $dx$) into two factors.
2. Let the factor without $dx$ equal $dv$.
3. Differentiate $u$ to find $du$ and integrate $dv$ to find $v$.
4. Use the formula $\int u\,du = uv - \int v\,du$.
5. Evaluate the right side of this equation to solve the integral.

### COMPOUND FUNCTIONS WHERE THE INNER FUNCTION IS AX

This shortcut works for compositions of functions $f(g(x))$ for which

# Empirical Model Building

- We know how to integrate the outer function $f$.
- The inner function $g(x)$ is of the form $ax$, which differentiates to a constant. Examples are as follows:

$$\int e^{2x}\,dx = \frac{1}{2}e^{2x} + C$$

$$\int \cos 3x\,dx = \frac{1}{3}\sin 3x + C$$

$$\int \tan 4x\,dx = \frac{1}{4}\sec^2 4x + C$$

## INTEGRATION BY PARTS

Integration by parts or partial integration is a process that finds the integral of a function that is a product of smaller functions. This is done in terms of the integral of the product of the smaller functions' derivative and antiderivative. Many methods have evolved over the years to executing integration by parts. One method is the "DI-agonal method." The basic form of integration by parts is presented below:

$$\int u\,dv = uv - \int v\,du$$

Examples are as follows:

$$\int x \ln x\,dx$$

$$\int u\,du = uv - \int v\,du$$

Let $u = \ln x$, $du = \frac{1}{x}dx$, $v = \frac{x^2}{2}$, $dv = x\,dx$

$$\int x \ln x\,dx = \frac{x^2}{2}\ln x - \int \frac{x^2}{2} \cdot \frac{1}{x}\,dx$$

$$= \frac{x^2}{2}\ln x - \int \frac{x}{2}\,dx$$

$$= \frac{x^2}{2}\ln x - \frac{x^2}{4} + C$$

$$\int \frac{x^3}{\sqrt{1-x^2}}\,dx$$

Let $u = -\dfrac{x^2}{2}$, $du = -xdx$, $dv = \dfrac{-2x}{\sqrt{1-x^2}}\,dx$

$$v = 2\sqrt{1-x^2}$$

$$\int \frac{x^3}{\sqrt{1-x^2}}\,dx = -x^2\sqrt{1-x^2} - \int (-2x)\sqrt{1-x^2}\,dx$$

$$= -x^2\sqrt{1+x^2} - \frac{2}{3}(1-x^2)^{\frac{2}{3}} + C$$

Readers are referred to standard calculus references for additional methods and examples of integration by parts.

## SYSTEMS MODELING FOR DATA ANALYTICS

Data analytics can be most useful if the data environment is viewed as a system. Many operational and functional elements interact to generate whatever data we are interested in. In a conventional definition, a system is defined as a collection of interrelated elements, whose collective and composite output, together, is higher than the mere sum of the outputs of the individual elements. With this viewpoint, each element in the system is recognized as a key driver or cornerstone in the overall system. Systems integration makes the world run smoothly for everyone. The output of data analytics must enmesh with the objectives, goals, and/or priorities of the organization. For this reason, data integration is as important as data collection. With today's interconnected world, as we have seen in the case of COVID-19, when something flares up in one corner of the world, it can quickly spread to other parts. Systems thinking enhances the functional interfaces as we endeavor to provide, not just scholarly insights into world developments but also social appreciation of what the society needs in terms of healthcare, education, mentoring, culture, diversity, work climate, gender equity, job training, leadership, respect, appreciation, recognition, reward, work compensation, digital work environment, career advancement opportunities, hierarchy of needs, and other dimensions of the work environment. There are several moving parts in the workplace in business, industry, government, academia, and the military. Only a systems view can ensure that all components are factored into the overall end goal. A systems view of data analytics allows an integrated design, analysis, and implementation of strategic plans. It would not work to have one segment of the enterprise embarking on one strategic approach, while another segment embraces practices that impede the overall achievement of an integrated organizational pursuit. In the context of operating in the global environment, whether a process is repeatable or not, in a statistical sense, is an issue of business stability and sustainment. A systems-based framework allows us to plan for prudent utilization of scarce human resources across all operations, particularly in response to developments, such as the COVID-19 pandemic. For the purpose of systems implementation for

# Empirical Model Building

data analytics, several systems models are available for consideration. These include the V-model, the waterfall model, the spiral model, the walking skeleton model, and the DEJI (Design, Evaluation, Justification, and Integration) systems model (Badiru, 2019). For the purpose of data integration, the DEJI model has a fitting applicability because of its steps of design, evaluation, justification, and integration. Figure 2.3 illustrates the basic structure and contents of the model. The DEJI model of systems engineering provides one additional option for systems engineering development applications. Although the DEJI model is generally applicable in all types of systems modeling, it is particularly well suited for data analytics. The core stages of the DEJI model applied to data analytics are as follows:

- Design of the data format
- Evaluation of the data elements and characteristics
- Justification of the data protocol
- Integration of the data analytics output into the prevailing operation scenario

Design encompasses any system initiative providing a starting point for a project. Thus, design can include technical product design, data format design, process initiation, and concept development. In essence, we can say that "design" represents requirements and specifications. Evaluation can use a variety of metrics both qualitative and quantitative, depending on the organization's needs. Justification can be done on the basis of monetary, technical, or social reasons. Not everything that is feasible is practical and desirable. Thus, justification is essential in the process of overall system design. After design, evaluation, and justification, integration needs to be done with respect to the normal or standard operations of the organization. Thus,

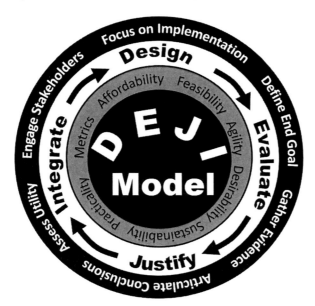

**FIGURE 2.3** The DEJI model for systems design, evaluation, justification, and integration.

the DEJI model provides an avenue of closing the loop in any system modeling challenge. This is especially useful for data analytics applications, where the end goal is expected to justify the means and analysis performed.

Empirical model building requires a structured approach executed in a sequence of questions related to what, when, where, who, how, and why. Some examples are summarized below:

Why is data analytics needed?
Who are the stakeholders?
What data is needed?
Where is the data available?
Who owns the data?
Who will collect the data?
How will the data be processed?
Who will perform the analytics?
How much time is needed for the process?

Along with the preceding discussions, communication, cooperation, and coordination are important for making the best use of the outputs of data analytics. Beyond the technical aspects of data analytics, the human aspects also play a crucial role. The nuances of human-system integration make it imperative that we also consider human needs for communication, cooperation, and coordination. The Triple C model presented by Badiru (2008) is an effective project planning and control tool. The model states that project management can be enhanced by implementing it within the integrated functions summarized below:

- Communication
- Cooperation
- Coordination

The model facilitates a systematic approach to project planning, organizing, scheduling, and control. The triple C model is distinguished from the 3C approach commonly used in military operations. The military approach emphasizes personnel management in the hierarchy of command, control, and communication. This places communication as the last function. The triple C model, by contrast, suggests communication as the first and foremost function. The triple C model can be implemented for project planning, scheduling, and control purposes. The basic questions of what, who, why, how, where, and when revolve around the triple C model. It highlights what must be done and when. It can also help to identify the resources (personnel, equipment, facilities, etc.) required for each effort. It points out important questions such as

- Does each project participant know what the objective is?
- Does each participant know his or her role in achieving the objective?
- What obstacles may prevent a participant from playing his or her role effectively?

Empirical Model Building

Triple C can mitigate disparity between idea and practice because it explicitly solicits information about the critical aspects of a project in terms of the following queries:

*Types of communication*

- Verbal
- Written
- Body language
- Visual tools (e.g., graphical tools)
- Sensual (use of all five senses: sight, smell, touch, taste, and hearing – olfactory, tactile, auditory)
- Simplex (unidirectional)
- Half-duplex (bidirectional with time lag)
- Full-duplex (real-time dialogue)
- One-on-one
- One-to-many
- Many-to-one

*Types of cooperation*

- Proximity
- Functional
- Professional
- Social
- Romantic
- Power influence
- Authority influence
- Hierarchical
- Lateral
- Cooperation by intimidation
- Cooperation by enticement

*Types of coordination*

- Teaming
- Delegation
- Supervision
- Partnership
- Token-passing
- Baton hand-off

## TRIPLE C QUESTIONS

Questioning is the best approach to getting information for effective project management. Everything should be questioned. By up-front questions, we can preempt and avert project problems later on. Typical questions to ask under triple C approach are as follows:

- What is the purpose of the project?
- Who is in charge of the project?
- Why is the project needed?
- Where is the project located?
- When will the project be carried out?
- How will the project contribute to increased opportunities for the organization?
- What is the project designed to achieve?
- How will the project affect different groups of people within the organization?
- What will be the project approach or methodology?
- What other groups or organizations will be involved (if any)?
- What will happen at the end of the project?
- How will the project be tracked, monitored, evaluated, and reported?
- What resources are required?
- What are the associated costs of the required resources?
- How do the project objectives fit the goal of the organization?
- What respective contribution is expected from each participant?
- What level of cooperation is expected from each group?
- Where is the coordinating point for the project?

The key to getting everyone on board with a project is to ensure that task objectives are clear and comply with the principle of **SMART** as outlined below:

**Specific.** Task objective must be specific.
**Measurable.** Task objective must be measurable.
**Aligned.** Task objective must be achievable and aligned with overall project goal.
**Realistic**. Task objective must be realistic and relevant to the organization.
**Timed.** Task objective must have a time basis.

If a task has the above intrinsic characteristics, then the function of communicating the task will more likely lead to personnel cooperation.

## COMMUNICATION

Communication makes working together possible. The communication function of project management involves making all those concerned become aware of project requirements and progress. Those who will be affected by the project directly or indirectly, as direct participants or as beneficiaries, should be informed as appropriate regarding the following:

- Scope of the project
- Personnel contribution required

- Expected cost and merits of the project
- Project organization and implementation plan
- Potential adverse effects if the project should fail
- Alternatives, if any, for achieving the project goal
- Potential direct and indirect benefits of the project

The communication channel must be kept open throughout the project life cycle. In addition to internal communication, appropriate external sources should also be consulted. The project manager must

- Exude commitment to the project.
- Utilize the communication responsibility matrix.
- Facilitate multi-channel communication interfaces.
- Identify internal and external communication needs.
- Resolve organizational and communication hierarchies.
- Encourage both formal and informal communication links.

When clear communication is maintained between management and employees and among peers, many project problems can be averted. Project communication may be carried out in one or more of the following formats:

- One-to-many
- One-to-one
- Many-to-one
- Written and formal
- Written and informal
- Oral and formal
- Oral and informal
- Nonverbal gestures

Good communication is affected when what is implied is perceived as intended. Effective communications are vital to the success of any project. Despite the awareness that proper communications form the blueprint for project success, many organizations still fail in their communications functions. The study of communication is complex. Factors that influence the effectiveness of communication within a project organization structure include the following.

1. **Personal perception**. Each person perceives events on the basis of personal psychological, social, cultural, and experimental background. As a result, no two people can interpret a given event the same way. The nature of events is not always the critical aspect of a problem situation. Rather, the problem is often the different perceptions of the different people involved.
2. **Psychological profile**. The psychological makeup of each person determines personal reactions to events or words. Thus, individual needs and level of thinking will dictate how a message is interpreted.

3. **Social environment**. Communication problems sometimes arise because people have been conditioned by their prevailing social environment to interpret certain things in unique ways. Vocabulary, idioms, organizational status, social stereotypes, and economic situation are among the social factors that can thwart effective communication.
4. **Cultural background**. Cultural differences are among the most pervasive barriers to project communications, especially in today's multinational organizations. Language and cultural idiosyncrasies often determine how communication is approached and interpreted.
5. **Semantic and syntactic factors**. Semantic and syntactic barriers to communications usually occur in written documents. Semantic factors are those that relate to the intrinsic knowledge of the subject of the communication. Syntactic factors are those that relate to the form in which the communication is presented. The problems created by these factors become acute in situations where response, feedback, or reaction to the communication cannot be observed.
6. **Organizational structure**. Frequently, the organization structure in which a project is conducted has a direct influence on the flow of information and, consequently, on the effectiveness of communication. Organization hierarchy may determine how different personnel levels perceive a given communication.
7. **Communication media**. The method of transmitting a message may also affect the value ascribed to the message and consequently how it is interpreted or used. The common barriers to project communications are as follows:

- Inattentiveness
- Lack of organization
- Outstanding grudges
- Preconceived notions
- Ambiguous presentation
- Emotions and sentiments
- Lack of communication feedback
- Sloppy and unprofessional presentation
- Lack of confidence in the communicator
- Lack of confidence by the communicator
- Low credibility of communicator
- Unnecessary technical jargon
- Too many people involved
- Untimely communication
- Arrogance or imposition
- Lack of focus

Some suggestions on improving the effectiveness of communication are presented next. The recommendations may be implemented as appropriate for any of the forms

# Empirical Model Building

of communications listed earlier. The recommendations are for both the communicator and the audience.

1. Never assume that the integrity of the information sent will be preserved as the information passes through several communication channels. Information is generally filtered, condensed, or expanded by the receivers before relaying it to the next destination. When preparing a communication that needs to pass through several organization structures, one safeguard is to compose the original information in a concise form to minimize the need for recomposition of the project structure.
2. Give the audience a central role in the discussion. A leading role can help make a person feel a part of the project effort and responsible for the projects' success. He or she can then have a more constructive view of project communication.
3. Do homework and think through the intended accomplishment of the communication. This helps eliminate trivial and inconsequential communication efforts.
4. Carefully plan the organization of the ideas embodied in the communication. Use indexing or points of reference whenever possible. Grouping ideas into related chunks of information can be particularly effective. Present the short messages first. Short messages help create focus, maintain interest, and prepare the mind for the longer messages to follow.
5. Highlight why the communication is of interest and how it is intended to be used. Full attention should be given to the content of the message with regard to the prevailing project situation.
6. Elicit the support of those around you by integrating their ideas into the communication. The more people feel they have contributed to the issue, the more expeditious they are in soliciting the cooperation of others. The effect of the multiplicative rule can quickly garner support for the communication purpose.
7. Be responsive to the feelings of others. It takes two to communicate. Anticipate and appreciate the reactions of members of the audience. Recognize their operational circumstances and present your message in a form they can relate to.
8. Accept constructive criticism. Nobody is infallible. Use criticism as a springboard to higher communication performance.
9. Exhibit interest in the issue in order to arouse the interest of your audience. Avoid delivering your messages as a matter of a routine organizational requirement.
10. Obtain and furnish feedback promptly. Clarify vague points with examples.
11. Communicate at the appropriate time, at the right place, to the right people.
12. Reinforce words with positive action. Never promise what cannot be delivered. Value your credibility.
13. Maintain eye contact in oral communication and read the facial expressions of your audience to obtain real-time feedback.

14. Concentrate on listening as much as speaking. Evaluate both the implicit and explicit meanings of statements.
15. Document communication transactions for future references.
16. Avoid asking questions that can be answered yes or no. Use relevant questions to focus the attention of the audience. Use questions that make people reflect upon their words, such as "How do you think this will work?" compared to "Do you this will work?"
17. Avoid patronizing the audience. Respect their judgment and knowledge.
18. Speak and write in a controlled tempo. Avoid emotionally charged voice inflections.
19. Create an atmosphere for formal and informal exchange of ideas.
20. Summarize the objectives of the communication and how they will be achieved.

A communication responsibility matrix shows the linking of sources of communication and targets of communication. Cells within the matrix indicate the subject of the desired communication. There should be at least one filled cell in each row and each column of the matrix. This assures that each individual of a department has at least one communication source or target associated with him or her. With a communication responsibility matrix, a clear understanding of what needs to be communicated to whom can be developed. Communication in a project environment can take any of several forms. The specific needs of a project may dictate the most appropriate mode. Three popular computer communication modes are discussed next in the context of communicating data and information for project management.

**Simplex communication.** This is a unidirectional communication arrangement in which one project entity initiates communication to another entity or individual within the project environment. The entity addressed in the communication does not have mechanism or capability for responding to the communication. An extreme example of this is a one-way, top-down communication from top management to the project personnel. In this case, the personnel have no communication access or input to top management. A budget-related example is a case where top management allocates budget to a project without requesting and reviewing the actual needs of the project. Simplex communication is common in authoritarian organizations.

**Half-duplex communication.** This is a bidirectional communication arrangement whereby one project entity can communicate with another entity and receive a response within a certain time lag. Both entities can communicate with each other but not at the same time. An example of half-duplex communication is a project organization that permits communication with top management without a direct meeting. Each communicator must wait for a response from the target of the communication. A request and allocation without a budget meeting is another example of half-duplex data communication in project management.

**Full-duplex communication.** This involves a communication arrangement that permits a dialogue between the communicating entities. Both individuals and entities can communicate with each other at the same time or face-to-face. As long as there is no clash of words, this appears to be the most receptive communication mode. It allows participative project planning, in which each project participant has an opportunity to contribute to the planning process.

Each member of a project team needs to recognize the nature of the prevailing communication mode in the project. Management must evaluate the prevailing communication structure and attempt to modify it, if necessary, to enhance project functions. An evaluation of who is to communicate with whom about what may help improve the project data/information communication process. A communication matrix may include notations about the desired modes of communication between individuals and groups in the project environment.

## COOPERATION

The cooperation of the project personnel must be explicitly elicited. Merely voicing consent for a project is not enough assurance of full cooperation. The participants and beneficiaries of the project must be convinced of the merits of the project. Some of the factors that influence cooperation in a project environment include personnel requirements, resource requirements, budget limitations, past experiences, conflicting priorities, and lack of uniform organizational support. A structured approach to seeking cooperation should clarify the following:

- Cooperative efforts required
- Precedents for future projects
- Implication of lack of cooperation
- Criticality of cooperation to project success
- Organizational impact of cooperation
- Time frame involved in the project
- Rewards of good cooperation

Cooperation is a basic virtue of human interaction. More projects fail due to a lack of cooperation and commitment than any other project factors. To secure and retain the cooperation of project participants, you must elicit a positive first reaction to the project. The most positive aspects of a project should be the first items of project communication. For project management, there are different types of cooperation that should be understood.

**Functional cooperation.** This is the type of cooperation induced by the nature of the functional relationship between two groups. The two groups may be required to perform related functions that can only be accomplished through mutual cooperation.

**Social cooperation.** This is the type of cooperation effected by the social relationship between two groups. The prevailing social relationship motivates cooperation that may be useful in getting project work done.

**Legal cooperation.** This is the type of cooperation that is imposed through some authoritative requirement. In this case, the participants may have no choice other than to cooperate.

**Administrative cooperation.** This is the type of cooperation brought on by administrative requirements that make it imperative that two groups work together on a common goal.

**Associative cooperation.** This type of cooperation may also be referred to as collegiality. The level of cooperation is determined by the association that exists between two groups.

**Proximity cooperation.** Cooperation due to the fact that two groups are geographically close is referred to as proximity cooperation. Being close makes it imperative that the two groups work together.

**Dependency cooperation.** This is the type of cooperation caused by the fact that one group depends on another group for some important aspect. Such dependency is usually of a mutual two-way nature. One group depends on the other for one thing while the latter group depends on the former for some other thing.

**Imposed cooperation.** In this type of cooperation, external agents must be employed to induced cooperation between two groups. This is applicable for cases where the two groups have no natural reason to cooperate. This is where the approaches presented earlier for seeking cooperation can became very useful.

**Lateral cooperation.** This cooperation involves cooperation with peers and immediate associates. It is often easy to achieve because existing lateral relationships create an environment that is conducive for project cooperation.

**Vertical cooperation.** Vertical or hierarchical cooperation refers to cooperation that is implied by the hierarchical structure of the project. For example, subordinates are expected to cooperate with their vertical superiors.

Whichever type of cooperation is available in a project environment; the cooperative forces should be channeled toward achieving project goals. Documentation of the prevailing level of cooperation is useful for winning further support for a project. Clarification of project priorities will facilitate personnel cooperation. Relative priorities of multiple projects should be specified so that a priority to all groups within the organization. Some guidelines for securing cooperation for most projects are as follows:

- Establish achievable goals for the project.
- Clearly outline the individual commitments required.
- Integrate project priorities with existing priorities.
- Eliminate the fear of job loss due to industrialization.
- Anticipate and eliminate potential sources of conflict.
- Use an open-door policy to address project grievances.
- Remove skepticism by documenting the merits of the project.

**Commitment.** Cooperation must be supported with commitment. To cooperate is to support the ideas of a project. To commit is to willingly and actively participate in project efforts again and again through the thick and thin of the project. Provision of resources is one way that management can express commitment to a project. Success can be assured if personal commitment is coupled with the triple C model.

## COORDINATION

After the communication and cooperation functions have successfully been initiated, the efforts of the project personnel must be coordinated. Coordination facilitates harmonious organization of project efforts. The construction of a responsibility chart can be very helpful at this stage. A responsibility chart is a matrix consisting of columns of individual or functional departments and rows of required actions. Cells within the matrix are filled with relationship codes that indicate who is responsible for what. The matrix helps avoid neglecting crucial communication requirements and obligations. It can help resolve questions such as the following:

- Who is to do what?
- How long will it take?
- Who is to inform whom of what?
- Whose approval is needed for what?
- Who is responsible for which results?
- What personnel interfaces are required?
- What support is needed from whom and when?

## CONFLICT RESOLUTION IN DATA ANALYTICS

Conflicts can and do develop in any work environment. Conflicts, whether intended or inadvertent, prevents an organization from getting the most out of the work force. When implemented as an integrated process, the triple C model can help avoid conflicts in a project. When conflicts do develop, it can help in resolving the conflicts. The key to conflict resolution is open and direct communication, mutual cooperation, and sustainable coordination. Several sources of conflicts can exist in projects. Some of these are discussed below.

**Schedule conflict.** Conflicts can develop because of improper timing or sequencing of project tasks. This is particularly common in large multiple projects. Procrastination can lead to having too much to do at once, thereby creating a clash of project functions and discord among project team members. Inaccurate estimates of time requirements may lead to infeasible activity schedules. Project coordination can help avoid schedule conflicts.

**Cost conflict.** Project cost may not be generally acceptable to the clients of a project. This will lead to project conflict. Even if the initial cost of the project is acceptable, a lack of cost control during implementation can lead to conflicts. Poor budget allocation approaches and the lack of a financial feasibility study will cause cost conflicts later on in a project. Communication and coordination can help prevent most of the adverse effects of cost conflicts.

**Performance conflict.** If clear performance requirements are not established, performance conflicts will develop. Lack of clearly defined performance standards can lead each person to evaluate his or her own performance based on personal value judgments. In order to uniformly evaluate quality of work and monitor project progress, performance standards should be established by using the triple C approach.

**Management conflict.** There must be a two-way alliance between management and the project team. The views of management should be understood by the team. The views of the team should be appreciated by management. If this does not happen, management conflicts will develop. A lack of a two-way interaction can lead to strikes and industrial actions, which can be detrimental to project objectives. The triple C approach can help create a conducive dialogue environment between management and the project team.

**Technical conflict.** If the technical basis of a project is not sound, technical conflict will develop. New industrial projects are particularly prone to technical conflicts because of their significant dependence on technology. Lack of a comprehensive technical feasibility study will lead to technical conflicts. Performance requirements and systems specifications can be integrated through the triple C approach to avoid technical conflicts.

**Priority conflict.** Priority conflicts can develop if project objectives are not defined properly and applied uniformly across a project. Lack of a direct project definition can lead each project member to define his or her own goals which may be in conflict with the intended goal of a project. Lack of consistency of the project mission is another potential source of conflicts in priorities. Over-assignment of responsibilities with no guidelines for relative significance levels can also lead to priority conflicts. Communication can help defuse priority conflict.

**Resource conflict.** Resource allocation problems are a major source of conflict in project management. Competition for resources, including personnel, tools, hardware, and software, can lead to disruptive clashes among project members. The triple C approach can help secure resource cooperation.

**Power conflict.** Project politics lead to a power play which can adversely affect the progress of a project. Project authority and project power should be clearly delineated. Project authority is the control that a person has by virtue of his or her functional post. Project power relates to the clout and influence, which a person can exercise due to connections within the administrative structure. People with popular personalities can often wield a lot of project power in spite of low or nonexistent project authority. The triple C model can facilitate a positive marriage of project authority and power to the benefit of project goals. This will help define clear leadership for a project.

**Personality conflict.** Personality conflict is a common problem in projects involving a large group of people. The larger the project, the larger the size of the management team needed to keep things running. Unfortunately, the larger management team creates an opportunity for personality conflicts.

Communication and cooperation can help defuse personality conflicts. In summary, conflict resolution through triple C can be achieved by observing the following guidelines:
1. Confront the conflict and identify the underlying causes.
2. Be cooperative and receptive to negotiation as a mechanism for resolving conflicts.
3. Distinguish between proactive, inactive, and reactive behaviors in a conflict situation.
4. Use communication to defuse internal strife and competition.
5. Recognize that short-term compromise can lead to long-term gains.
6. Use coordination to work toward a unified goal.
7. Use communication and cooperation to turn a competitor into a collaborator.

It is the little and often neglected aspects of a project that lead to project failures. Several factors may constrain the project implementation. All the relevant factors can be evaluated under the triple C model right from the project initiation stage.

## REFERENCES

Badiru, Adedeji B. (2019), *Systems Engineering Models: Theory, Methods, and Applications*, Boca Raton, FL: Taylor & Francis Group/CRC Press.

Badiru, Adedeji B. (2008), *Triple C Model of Project Management: Communication, Cooperation, and Coordination*, Boca Raton, FL: Taylor & Francis Group/CRC Press.

# 3 Data Visualization Methods

Data viewed is data appreciated.

## INTRODUCTION TO DATA VISUALIZATION

Statistical data management is essential for measurement with respect to analyzing and interpreting measurement outputs. In this chapter, a project control scenario is used to illustrate data management for measurement of project performance. The data presentation techniques presented in this chapter are translatable to other data analytics platforms. The present age of computer software, hardware, and tools offers a vast array of techniques for data visualization, beyond what is presented in this chapter. Readers are encouraged to refer to the latest commercial and open-source software for data visualization. More important, the prevalence of cloud-based subscription software products can assist with on-demand data visualization needs. Those online tools should be leveraged at the time of need. The chapter presents only basic and standard methods to spark and guide the interest and awareness of readers.

## CASE EXAMPLE OF "COVIDVISUALIZER" WEBSITE

For challenges of interest, such as the COVID-19 pandemic, data visualization can generate an immediate impact of understanding and appreciation, and, consequently, the determination of the lines of action needed. Tracking the fast worldwide spread of coronavirus helped to heighten the necessity and utility of data visualization. In the wake of COVID-19, several online data visualization tools evolved quickly to inform and educate the public about the disease's spread. One of the earliest such tools was the *www.covidvisualizer.com* website, which was developed by **Navid Mamoon** and **Gabriel Rasskin**, two undergraduate students at **Carnegie Mellon University** in 2020. The goal of the project is to provide a simple interactive way to visualize the impact of COVID-19. The developers want people to be able to see the effort as something that brings people all together in the collective worldwide fight against COVID-19. The website has a colorful and visually pleasing (almost trance-inducing) rotation of the Earth. Clicking on a country as it rotates by bringing up the country's up-to-the-minute current statistics for COVID-19. The information displayed includes the following:

- Country name
- Country flag
- Total cases

- Active cases
- Deceased
- Recovered cases
- Line chart (trend line) over time for active, deaths, and recovered

In response to the developers' solicitation of questions, suggestions, or feedback, I had the pleasure of contacting them to offer the suggestion of adding a search tool to the website. The original website design only has access to each country's information only when it is clicked during the rotational cycle of geography without the benefit of having written names of the countries. This means that a user has to know which country is which on the world map in order to click on it. Unfortunately, not all users can identify specific countries on the world map. Further, some countries are so tiny that clicking on them on a rotating globe is practically impossible. The idea of a search tool is to improve the user-friendliness of the website by providing a way to search for a specific country of interest. The developers were excited about the feedback and implemented a by-name search tool. The confirmation of below (date March 28, 2020) is the response:

> Thank you for requesting the **search feature** on our website, *covidvisualizer.com*. We apologize for the delay, (it can take a while to develop a feature like this) but there is now a search function running on the site! You can search by country name or ISO code by simply clicking the new little search icon.

Unfortunately, within 24 hours, the search tool was removed, for which I reengaged with the developers. The response of March 29, 2020, is echoed below:

> We unfortunately disabled it, it caused some issues with our server and we'll have to develop it further.

Apparently, adding a search tool caused the website computer server to crash. The developers responded to the suggestion and they developed the visualization tool further. In a subsequent version of the website, the developers included two stable and sustainable search tools, through which a user can search by country name or by scrolling through the alphabetical listing of all countries. The website has enjoyed a consistent worldwide usage since it was introduced in early March 2020. I am delighted and proud that, from a user perspective, I was able to provide mentoring and technical feedback to the website developers. The lesson and moral of this center around the fact that we are all in the fight against COVID-19 together and teamwork is essential for success. In addition, user assessment and feedback are essential for product advancement regardless of whether the product is a commercial product or an open-source tool available free online. Thus, making a contribution to the utility of this very useful website is a proud accomplishment that bears out the theme of this chapter and the entire book.

## DYNAMISM AND VOLATILITY OF DATA

The data to be visually presented can be dynamic, volatile, and elusive. The more we can know about the characteristics of the data, the better we can design, evaluate, and implement the technical protocol to handle the data. *Transient data* is defined as a volatile set of data that is used for one-time decision-making and is not then needed again. An example may be the number of operators that show up at a job site on a given day. Unless there is some correlation between the day-to-day attendance records of operators, this piece of information will have relevance only for that given day. The project manager can make his decision for that day on the basis of that day's attendance record. Transient data need not be stored in a permanent database unless it may be needed for future analysis or uses (e.g., forecasting, incentive programs, performance review).

*Recurring data* refers to data that is encountered frequently enough to necessitate storage on a permanent basis. An example is a file containing contract due dates. This file will need to be kept at least through the project life cycle. Recurring data may be further categorized into *static data* and *dynamic data*. A recurring data that is static will retain its original parameters and values each time it is retrieved and used. A recurring data that is dynamic has the potential for taking on different parameters and values each time it is retrieved and used. Storage and retrieval considerations for project control should address the following questions:

1. What is the origin of the data?
2. How long will the data be maintained?
3. Who needs access to the data?
4. What will the data be used for?
5. How often will the data be needed?
6. Is the data for look-up purposes only (i.e., no printouts)?
7. Is the data for reporting purposes (i.e., generate reports)?
8. In what format is the data needed?
9. How fast will the data need to be retrieved?
10. What security measures are needed for the data?

## DATA DETERMINATION AND COLLECTION

It is essential to determine what data to collect for project control purposes. Data collection and analysis are the basic components of generating information for project control. The requirements for data collection are discussed next.

### Choosing the Data

This involves selecting data on the basis of their relevance and the level of likelihood that they will be needed for future decisions and whether or not they contribute to making the decision better. The intended users of the data should also be identified.

### Collecting the Data

This identifies a suitable method of collecting the data as well as the source from which the data will be collected. The collection method will depend on the particular operation being addressed. The common methods include manual tabulation, direct keyboard entry, optical character reader, magnetic coding, electronic scanner, and, more recently, voice command. An input control may be used to confirm the accuracy of collected data. Examples of items to control when collecting data are the following:

### Relevance Check

This checks if the data is relevant to the prevailing problem. For example, data collected on personnel productivity may not be relevant for a decision involving marketing strategies.

### Limit Check

This checks to ensure that the data is within known or acceptable limits. For example, an employee overtime claim amounting to over 80 hours per week for several weeks in a row is an indication of a record well beyond ordinary limits.

### Critical Value

This identifies a boundary point for data values. Values below or above a critical value fall in different data categories. For example, the lower specification limit for a given characteristic of a product is a critical value that determines whether or not the product meets quality requirements.

### Coding the Data

This refers to the technique used in representing data in a form useful for generating information. This should be done in a compact and yet meaningful format. The performance of information systems can be greatly improved if effective data formats and coding are designed into the system right from the beginning.

### Processing the Data

Data processing is the manipulation of data to generate useful information. Different types of information may be generated from a given data set depending on how it is processed. The processing method should consider how the information will be used, who will be using it, and what caliber of system response time is desired. If possible, processing controls should be used.

### Control Total

It checks the completeness of the processing by comparing accumulated results to a known total. An example of this is the comparison of machine throughput to a standard production level or the comparison of cumulative project budget depletion to a cost accounting standard.

## Consistency Check

It checks if the processing is producing the same results for similar data. For example, an electronic inspection device that suddenly shows a measurement that is ten times higher than the norm warrants an investigation of both the input and the processing mechanisms.

## Scales of Measurement

For numeric scales, specify units of measurement, increments, the zero point on the measurement scale, and the range of values.

## Using the Information

Using information involves people. Computers can collect data, manipulate data, and generate information, but the ultimate decision rests with people, and decision-making starts when information becomes available. Intuition, experience, training, interest, and ethics are just a few of the factors that determine how people use information. The same piece of information that is positively used to further the progress of a project in one instance may also be used negatively in another instance. To assure that data and information are used appropriately, computer-based security measures can be built into the information system. Project data may be obtained from several sources. Some potential sources are as follows:

- Formal reports
- Interviews and surveys
- Regular project meetings
- Personnel time cards or work schedules

The timing of data is also very important for project control purposes. The contents, level of detail, and frequency of data can affect the control process. An important aspect of project management is the determination of the data required to generate the information needed for project control. The function of keeping track of the vast quantity of rapidly changing and interrelated data about project attributes can be very complicated. The major steps involved in data analysis for project control are as follows:

- Data collection
- Data analysis and presentation
- Decision-making
- Implementation of action

Data is processed to generate information. Information is analyzed by the decision maker to make the required decisions. Good decisions are based on timely and relevant information, which in turn is based on reliable data. Data analysis for project control may involve the following functions:

- Organizing and printing computer-generated information in a form usable by managers
- Integrating different hardware and software systems to communicate in the same project environment
- Incorporating new technologies such as expert systems into data analysis
- Using graphics and other presentation techniques to convey project information

Proper data management will prevent misuse, misinterpretation, or mishandling. Data is needed at every stage in the life cycle of a project from the problem identification stage through the project phase-out stage. The various items for which data may be needed are project specifications, feasibility study, resource availability, staff size, schedule, project status, performance data, and phase-out plan. The documentation of data requirements should cover the following:

- **Data summary**. A data summary is a general summary of the information and decision for which the data is required as well as the form in which the data should be prepared. The summary indicates the impact of the data requirements on the organizational goals.
- **Data processing environment**. The processing environment identifies the project for which the data is required, the user personnel, and the computer system to be used in processing the data. It refers to the project request or authorization and relationship to other projects and specifies the expected data communication needs and mode of transmission.
- **Data policies and procedures**. Data handling policies and procedures describe policies governing data handling, storage, and modification and the specific procedures for implementing changes to the data. Additionally, they provide instructions for data collection and organization.
- **Static data**. A static data description describes that portion of the data that is used mainly for reference purposes and it is rarely updated.
- **Dynamic data**. A dynamic data description describes that portion of the data that is frequently updated based on the prevailing circumstances in the organization.
- **Data frequency**. The frequency of data update specifies the expected frequency of data change for the dynamic portion of the data, for example, quarterly. This data change frequency should be described in relation to the frequency of processing.
- **Data constraints.** Data constraints refer to the limitations on the data requirements. Constraints may be procedural (e.g., based on corporate policy), technical (e.g., based on computer limitations), or imposed (e.g., based on project goals).
- **Data compatibility.** Data compatibility analysis involves ensuring that data collected for project control needs will be compatible with future needs.

# Data Visualization Methods

- **Data contingency.** A data contingency plan concerns data security measures in case of accidental or deliberate damage or sabotage affecting hardware, software, or personnel.

## DATA EXPLOITATION

Data availability should be exploited and leverage for pertinent decision-making. Data exploitation refers to the various mathematical and graphical operations that can be performed on data to elicit the inherent information contained in the data. The manner in which project data is analyzed and presented can affect how the information is perceived by the decision maker. The examples presented in this section illustrate how basic data analysis techniques can be used to convey important information for project control.

In many cases, data is represented as the answer to direct questions such as the following: When is the project deadline? Who are the people assigned to the first task? How many resource units are available? Are enough funds available for the project? What are the quarterly expenditures on the project for the past two years? Is personnel productivity low, average, or high? Who is the person in charge of the project? Answers to these types of questions constitute data of different forms or expressed on different scales. The resulting data may be qualitative or quantitative. Different techniques are available for analyzing the different types of data. This section discusses some of the basic techniques for data analysis. The data presented in Table 3.1 is used to illustrate the data analysis techniques.

### Raw Data

Raw data consists of ordinary observations recorded for a decision variable or factor. Examples of factors for which data may be collected for decision-making are revenue, cost, personnel productivity, task duration, project completion time, product quality, and resource availability. Raw data should be organized into a format suitable for visual review and computational analysis. The data in Table 3.1 represents the quarterly revenues from projects A, B, C, and D. For example, the data for quarter 1 indicates that project C yielded the highest revenue of $4,500,000, while project B yielded the lowest revenue of $1,200,000. Figure 3.1 presents the raw data of project revenue as a line graph. The same information is presented as a multiple bar chart in Figure 3.2.

### TABLE 3.1
### Quarterly Revenue from Four Projects (in $1,000s)

| Project | Quarter 1 | Quarter 2 | Quarter 3 | Quarter 4 | Row Total |
|---------|-----------|-----------|-----------|-----------|-----------|
| A       | 3,000     | 3,200     | 3,400     | 2,800     | 12,400    |
| B       | 1,200     | 1,900     | 2,500     | 2,400     | 8,000     |
| C       | 4,500     | 3,400     | 4,600     | 4,200     | 16,700    |
| D       | 2,000     | 2,500     | 3,200     | 2,600     | 10,300    |
| Total   | 10,700    | 11,000    | 13,700    | 12,000    | 47,400    |

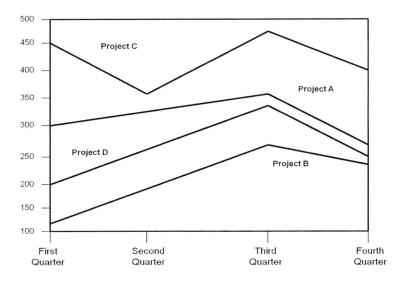

**FIGURE 3.1**  Line graph of quarterly project revenues.

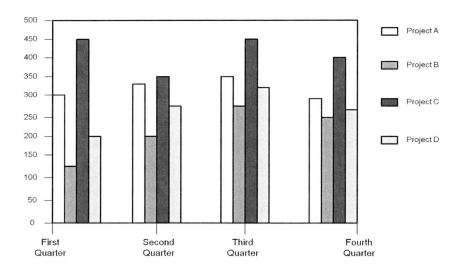

**FIGURE 3.2**  Multiple bar chart of quarterly project revenues.

## Total Revenue

A total or sum is a measure that indicates the overall effect of a particular variable. If $X_1, X_2, X_3, ..., X_n$ represent a set of $n$ observations (e.g., revenues), then the total is computed as follows:

$$T = \sum_{i=1}^{n} X_i$$

## Data Visualization Methods

For the data in Table 3.1, the total revenue for each project is shown in the last column. The totals indicate that project C brought in the largest total revenue over the four quarters under consideration, while project B produced the lowest total revenue. The last row of the table shows the total revenue for each quarter. The totals reveal that the largest revenue occurred in the third quarter. The first quarter brought in the lowest total revenue. The grand total revenue for the four projects over the four quarters is shown as $47,400,000 in the last cell in the table. The total revenues for the four projects over the four quarters are shown in a pie chart in Figure 3.3. The percentage of the overall revenue contributed by each project is also shown on the pie chart.

### Average Revenue

Average is one of the most used measures in data analysis. Given $n$ observations (e.g., revenues), $X_1, X_2, X_3, \ldots, X_n$, the average of the observations is computed as

$$\bar{X} = \frac{\sum_{i=1}^{n} X_i}{n}$$

$$= \frac{T_x}{n}$$

where $T_x$ is the sum of $n$ revenues. For our sample data, the average quarterly revenues for the four projects are

$$\bar{X}_A = \frac{(3{,}000 + 3{,}200 + 3{,}400 + 2{,}800)(\$1{,}000)}{4}$$

$$= \$3{,}100{,}000$$

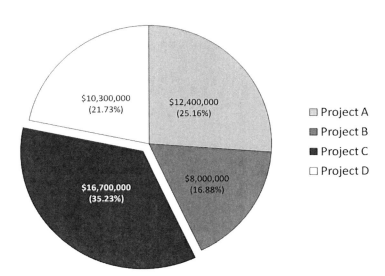

**FIGURE 3.3** Pie chart of total revenue per project.

$$\bar{X}_B = \frac{(1{,}200 + 1{,}900 + 2{,}500 + 2{,}400)(\$1{,}000)}{4}$$

$$= \$2{,}000{,}000$$

$$\bar{X}_C = \frac{(4{,}500 + 3{,}400 + 4{,}600 + 4{,}200)(\$1{,}000)}{4}$$

$$= \$4{,}175{,}000$$

$$\bar{X}_D = \frac{(2{,}000 + 2{,}500 + 3{,}200 + 2{,}600)(\$1{,}000)}{4}$$

$$= 2{,}575{,}000$$

Similarly, the expected average revenues per project for the four quarters are

$$\bar{X}_1 = \frac{(3{,}000 + 1{,}200 + 4{,}500 + 2{,}000)(\$1{,}000)}{4}$$

$$= \$2{,}675{,}000$$

$$\bar{X}_2 = \frac{(3{,}200 + 1{,}900 + 3{,}400 + 2{,}500)(\$1{,}000)}{4}$$

$$= \$2{,}750{,}000$$

$$\bar{X}_3 = \frac{(3{,}400 + 2{,}500 + 4{,}600 + 3{,}200)(\$1{,}000)}{4}$$

$$= \$3{,}425{,}000$$

$$\bar{X}_4 = \frac{(2{,}800 + 2{,}400 + 4{,}200 + 2{,}600)(\$1{,}000)}{4}$$

$$= \$3{,}000{,}000$$

The above values are shown in a bar chart in Figure 3.4. The average revenue from any of the four projects in any given quarter is calculated as the sum of all the observations divided by the number of observations. That is,

$$\bar{\bar{X}} = \frac{\sum_{i=1}^{N} \sum_{j=1}^{M} X_{ij}}{K}$$

where

$N$ is the number of projects.
$M$ is the number of quarters.
$K$ is the total number of observations ($K = NM$).

# Data Visualization Methods

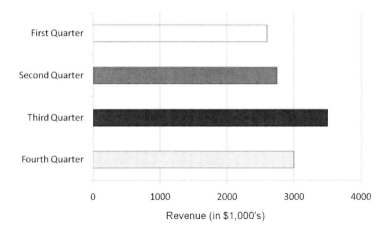

**FIGURE 3.4** Average revenue per project for each quarter.

The overall average per project per quarter is

$$\overline{\overline{X}} = \frac{\$47,400,000}{16}$$

$$= \$2,962,500$$

As a cross-check, the sum of the quarterly averages should be equal to the sum of the project revenue averages, which is equal to the grand total divided by 4.

$$(2,675+2,750+3,425+3,000)(\$1,000) = (3,100+2,000+4,175+2,575)(\$1,000)$$

$$= \$11,800,000$$

$$= \$47,400,000/4$$

The cross-check procedure above works because we have a balanced table of observations. That is, we have four projects and four quarters. If there were only three projects, for example, the sum of the quarterly averages would not be equal to the sum of the project averages.

## Median Revenue

The median is the value that falls in the middle of a group of observations arranged in order of magnitude. One-half of the observations are above the median, and the other half are below the median. The method of determining the median depends on whether or not the observations are organized into a frequency distribution. For unorganized data, it is necessary to arrange the data in an increasing or decreasing order before finding the median. Given $K$ observations (e.g., revenues), $X_1, X_2, X_3, \ldots, X_K$, arranged in increasing or decreasing order, the median is identified as the value in position $(K+1)/2$ in the data arrangement if $K$ is an odd number. If $K$ is an even number, then the average of the two middle values is considered to be the median. If the sample data are arranged in increasing order, we would get the following:

1,200, 1,900, 2,000, 2,400, 2,500, 2,500, 2,600, 2,800, 3,000, 3,200, 3,200, 3,400, 3,400, 4,200, 4,500, and 4,600

The median is then calculated as (2,800+3,000)/2 = 2,900. Half of the recorded revenues are expected to be above $2,900,000, while half are expected to be below that amount. Figure 3.5 presents a bar chart of the revenue data arranged in increasing order. The median is anywhere between the eighth and ninth values in the ordered data.

## Quartiles and Percentiles

The median is a position measure because its value is based on its position in a set of observations. Other measures of position are *quartiles* and *percentiles*. There are three quartiles that divide a set of data into four equal categories. The first quartile, denoted $Q_1$, is the value below which one-fourth of all the observations in the data set fall. The second quartile, denoted $Q_2$, is the value below which two-fourths or one-half of all the observations in the data set fall. The third quartile, denoted $Q_3$, is the value below which three-fourths of the observations fall. The second quartile is identical to the median. It is technically incorrect to talk of the fourth quartile because it will imply that there is a point within the data set below which all the data points fall: a contradiction! A data point cannot lie within the range of the observations and at the same time exceed all the observations, including itself.

The concept of percentiles is similar to the concept of quartiles except that reference is made to percentage points. There are 99 percentiles that divide a set of observations into 100 equal parts. The $X$ percentile is the value below which $X$ percent

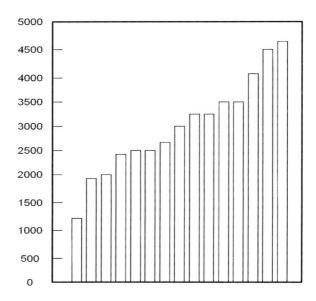

**FIGURE 3.5** Ordered bar chart.

of the data fall. The 99 percentile refers to the point below which 99 percent of the observations fall. The three quartiles discussed previously are regarded as the 25th, 50th, and 75th percentiles. It would be technically incorrect to talk of the 100 percentile. For the purpose of doing performance rating, such as on an examination or a product quality assessment, the higher the percentile of an individual or product, the better. In many cases, recorded data are classified into categories that are not indexed to numerical measures. In such cases, other measures of central tendency or position will be needed. An example of such a measure is the mode.

## The Mode

The mode is defined as the value that has the highest frequency in a set of observations. When the recorded observations can be classified only into categories, the mode can be particularly helpful in describing the data. Given a set of $K$ observations (e.g., revenues), $X_1, X_2, X_3, \ldots, X_K$, the mode is identified as that value that occurs more than any other value in the set. Sometimes, the mode is not unique in a set of observations. For example, in Table 3.2, $2,500, $3,200, and $3,400 all have the same number of occurrences. Each of them is a mode of the set of revenue observations. If there is a unique mode in a set of observations, then the data is said to be unimodal. The mode is very useful in expressing the central tendency for observations with qualitative characteristics such as color, marital status, or state of origin.

## Range of Revenue

The range is determined by the two extreme values in a set of observations. Given $K$ observations (e.g., revenues), $X_1, X_2, X_3, \ldots, X_K$, the range of the observations is simply the difference between the lowest and the highest observations. This measure is useful when the analyst wants to know the extent of extreme variations in a parameter. The range of the revenues in our sample data is ($4,600,000 − $1,200,000) = $3,400,000. Because of its dependence on only two values, the range tends to increase as the sample size increases. Furthermore, it does not provide a measurement of the variability of the observations relative to the center of the distribution. This is why the standard deviation is normally used as a more reliable measure of dispersion than the range.

The variability of a distribution is generally expressed in terms of the deviation of each observed value from the sample average. If the deviations are small, the set of data is said to have low variability. The deviations provide information about the degree of dispersion in a set of observations. A general formula to evaluate the variability of data cannot be based on the deviations. This is because some of the deviations are negative, whereas some are positive and the sum of all the deviations is equal to 0. One possible solution to this is to compute the average deviation.

## Average Deviation

The average deviation is the average of the absolute values of the deviations from the sample average. Given $K$ observations (e.g., revenues), $X_1, X_2, X_3, \ldots, X_K$, the average deviation of the data is computed as

$$\bar{D} = \frac{\sum_{i=1}^{K}|X_i - \bar{X}|}{K}$$

Table 3.2 shows how the average deviation is computed for our sample data. One aspect of the average deviation measure is that the procedure ignores the sign associated with each deviation. Despite this disadvantage, its simplicity and ease of computation make it useful. In addition, the knowledge of the average deviation helps in understanding the standard deviation, which is the most important measure of dispersion available.

## Sample Variance

Sample variance is the average of the squared deviations computed from a set of observations. If the variance of a set of observations is large, the data is said to have a large variability. For example, a large variability in the levels of productivity of a project team may indicate a lack of consistency or improper methods in the project functions. Given $K$ observations (e.g., revenues), $X_1, X_2, X_3, \ldots, X_K$, the sample variance of the data is computed as

$$s^2 = \frac{\sum_{i=1}^{K}(X_i - \bar{X})^2}{K-1}$$

**TABLE 3.2**
**Average Deviation, Standard Deviation, and Variance**

| Observation Number ($i$) | Recorded Observation $X_i$ | Deviation from Average $X_i - \bar{X}$ | Absolute Value $\|X_i - \bar{X}\|$ | Square of Deviation $(X_i - \bar{X})^2$ |
|---|---|---|---|---|
| 1 | 3,000 | 37.5 | 37.5 | 1,406.25 |
| 2 | 1,200 | −1,762.5 | 1762.5 | 3,106,406.30 |
| 3 | 4,500 | 1,537.5 | 1537.5 | 2,363,906.30 |
| 4 | 2,000 | −962.5 | 962.5 | 926,406.25 |
| 5 | 3,200 | 237.5 | 237.5 | 56,406.25 |
| 6 | 1,900 | −1,062.5 | 1062.5 | 1,128,906.30 |
| 7 | 3,400 | 437.5 | 437.5 | 191,406.25 |
| 8 | 2,500 | −462.5 | 462.5 | 213,906.25 |
| 9 | 3,400 | 437.5 | 437.5 | 191,406.25 |
| 10 | 2,500 | −462.5 | 462.5 | 213,906.25 |
| 11 | 4,600 | 1,637.5 | 1637.5 | 2,681,406.30 |
| 12 | 3,200 | 237.5 | 237.5 | 56,406.25 |
| 13 | 2,800 | −162.5 | 162.5 | 26,406.25 |
| 14 | 2,400 | −562.5 | 562.5 | 316,406.25 |
| 15 | 4,200 | 1,237.5 | 1237.5 | 1,531,406.30 |
| 16 | 2,600 | −362.5 | 362.5 | 131,406.25 |
| Total | 47,400.0 | 0.0 | 11,600.0 | 13,137,500.25 |
| Average | 2,962.5 | 0.0 | 725.0 | 821,093.77 |
| Square root | — | — | — | 906.14 |

# Data Visualization Methods

The variance can also be computed by the following alternate formulas:

$$s^2 = \frac{\sum_{i=1}^{K}\left(X_i^2 - \left(\frac{1}{K}\right)\right)\left[\sum_{i=1}^{K} X_i\right]^2}{K-1}$$

$$s^2 = \frac{\sum_{i=1}^{K} X_i^2 - K(\bar{X}^2)}{K-1}$$

Using the first formula, the sample variance of the data in Table 3.2 is calculated as

$$s^2 = \frac{13{,}137{,}500.25}{16-1}$$

$$= 875{,}833.33$$

The average calculated in the last column of Table 3.1 is obtained by dividing the total for that column by 16 instead of $16-1 = 15$. That average is not the correct value of the sample variance. However, as the number of observations gets very large, the average as computed in the table will become a close estimate for the correct sample variance. Analysts make a distinction between the two values by referring to the number calculated in the table as the population variance when $K$ is very large and referring to the number calculated by the formulas above as the sample variance particularly when $K$ is small. For our example, the population variance is given by

$$\sigma^2 = \frac{\sum_{i=1}^{K}(X_i - \bar{X})^2}{K}$$

$$= \frac{13{,}137{,}500.25}{16}$$

$$= 821{,}093.77$$

while the sample variance, as shown previously for the same data set, is given by

$$\sigma^2 = \frac{\sum_{i=1}^{K}(X_i - \bar{X})^2}{K-1}$$

$$= \frac{13{,}137{,}500.25}{(16-1)}$$

$$= 875{,}833.33$$

## Standard Deviation

The sample standard deviation of a set of observations is the positive square root of the sample variance. The use of variance as a measure of variability has some drawbacks. For example, the knowledge of the variance is helpful only when two or more sets of observations are compared. Because of the squaring operation, the variance is expressed in square units rather than the original units of the raw data. To get a reliable feel for the variability in the data, it is necessary to restore the original units by performing the square root operation on the variance. This is why standard deviation is a widely recognized measure of variability. Given $K$ observations (e.g., revenues), $X_1, X_2, X_3, \ldots, X_K$, the sample standard deviation of the data is computed as

$$s = \sqrt{\frac{\sum_{i=1}^{K}(X_i - \bar{X})^2}{K-1}}$$

As in the case of the sample variance, the sample standard deviation can also be computed by the following alternate formulas:

$$s = \sqrt{\frac{\sum_{i=1}^{K} X_i^2 - \left(\frac{1}{K}\right)\left[\sum_{i=1}^{K} X_i\right]^2}{K-1}}$$

$$s = \sqrt{\frac{\sum_{i=1}^{K} X_i^2 - K(\bar{X})^2}{K-1}}$$

Using the first formula, the sample standard deviation of the data is calculated as

$$s = \sqrt{\frac{13,137,500.25}{(16-1)}}$$

$$= \sqrt{875,833.33}$$

$$= 935.8597$$

We can say that the variability in the expected revenue per project per quarter is $935,859.70. The population sample standard deviation is given by the following:

$$\sigma = \sqrt{\frac{\sum_{i=1}^{K}(X_i - \bar{X})^2}{K}}$$

$$= \sqrt{\frac{13,137,500.25}{16}}$$

$$= \sqrt{821,093.77}$$

$$= 906.1423$$

The sample standard deviation is given by the following expression:

# Data Visualization Methods

$$s = \sqrt{\frac{\sum_{i=1}^{K}(X_i - \bar{X})^2}{K-1}}$$

$$= \sqrt{\frac{13{,}137{,}500.25}{(16-1)}}$$

$$= 935.8597$$

The results of data analysis can be reviewed directly to determine where and when project control actions may be needed. The results can also be used to generate control charts, as illustrated in Chapter 1 for my high school course grades.

# 4 Basic Mathematical Calculations for Data Analytics

Where there is no calculation, there is no compliance.

## INTRODUCTION TO CALCULATION FOR DATA ANALYTICS

Data analytics is, inherently, dependent on mathematical calculations. For ease of reference and use, a comprehensive collection of basic mathematical calculations is presented in this chapter. Illustrative and/or clarifying computational examples are provided for some of the equations and formulas. Readers are encouraged to go through the collection to find equations that may be applicable and useful for the data set of interest. Many times, seeing a presentation of a new equation may spark an idea of what type of data modeling technique is relevant for a data set.

## QUADRATIC EQUATION

$$ax^2 + bx + c = 0$$

Solution:

$$x = \frac{-b \pm \sqrt{b^2 - 4ac}}{2a}$$

If $b^2 - 4ac < 0$, the roots are complex.
If $b^2 - 4ac > 0$, the roots are real.
If $b^2 - 4ac = 0$, the roots are real and repeated.

Dividing both sides of Eq. (1.1) by "$a$" ($a \neq 0$),

$$x^2 + \frac{b}{a}x + \frac{c}{a} = 0$$

Note if $a = 0$, the solution to $ax^2 + bx + c = 0$ is $x = -\frac{c}{b}$.

Rewrite Eq. (1.1) as

$$\left(x + \frac{b}{2a}\right)^2 - \frac{b^2}{4a^2} + \frac{c}{a} = 0$$

$$\left(x+\frac{b}{2a}\right)^2 = \frac{b^2}{4a^2} - \frac{c}{a} = \frac{b^2-4ac}{4a^2}$$

$$x+\frac{b}{2a} = \pm\sqrt{\frac{b^2-4ac}{4a^2}} = \pm\frac{\sqrt{b^2-4ac}}{2a}$$

$$x = -\frac{b}{2a} \pm \sqrt{\frac{b^2-4ac}{4a^2}}$$

$$x = \frac{-b \pm \sqrt{b^2-4ac}}{2a}$$

## Overall Mean

$$\bar{x} = \frac{n_1\bar{x}_1 + n_2\bar{x}_2 + n_3\bar{x}_3 + \ldots + n_k\bar{x}_k}{n_1 + n_2 + n_3 + \ldots + n_k} = \frac{\sum n\bar{x}}{\sum n}$$

## Chebyshev's Theorem

$$1 - 1/k^2$$

## Permutations

A permutation of $m$ elements from a set of $n$ elements is any arrangement, without repetition, of $m$ elements. The total number of all the possible permutations of $n$ distinct objects taken $m$ times is

$$P(n,m) = \frac{n!}{(n-m)!} \quad (n \geq m)$$

**Example:**
Find the number of ways a president, vice president, secretary, and a treasurer can be chosen from a committee of eight members.

**Solution:**

$$P(n,m) = \frac{n!}{(n-m)!} = P(8,4) = \frac{8!}{(8-4)!} = \frac{8.7.6.5.4.3.2.1}{4.3.2.1} = 1,680$$

There are 1,680 ways of choosing the four officials from the committee of eight members.

## Combinations

The number of combinations of $n$ distinct elements taken is given by

$$C(n,m) = \frac{n!}{m!(n-m)!} \quad (n \geq m)$$

# Basic Mathematical Calculations

**Example:**
How many poker hands of five cards can be dealt from a standard deck of 52 cards?

**Solution:**
Note: The order in which the five cards are dealt is not important.

$$C(n,m) = \frac{n!}{m!(n-m)!} = C(52,5) = \frac{52!}{5!(52-5)!} = \frac{52!}{5!47!}$$

$$= \frac{52.51.50.49.48}{5.4.3.2.1} = 2,598,963$$

## FAILURE

$$q = 1 - p = \frac{n-s}{n}$$

## PROBABILITY DISTRIBUTION

An example of probability distribution is shown in the histogram plot in Figure 4.1. A probability density function can be inferred from the probability distribution as shown by the example in Figure 4.2.

## PROBABILITY

$$P(X \leq x) = F(x) = \int_{-\infty}^{x} f(x)dx$$

## DISTRIBUTION FUNCTION

Figure 4.3 shows the general profile of the cumulative probability function of the probability density function in Figure 4.2.

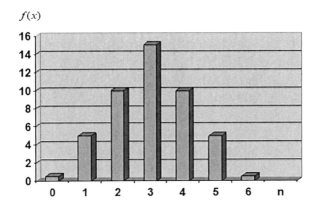

**FIGURE 4.1** Probability distribution plot.

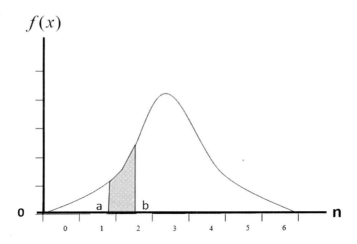

**FIGURE 4.2**   Example of probability density function.

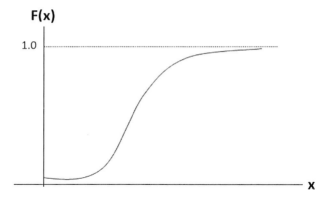

**FIGURE 4.3**   Cumulative probability plot.

## Expected Value

$$\mu = \sum (xf(x))$$

## Variance

$$\sigma^2 = \sum (x-\mu)^2 f(x) \quad \text{or} \quad \sigma^2 = \int_{-\infty}^{\infty} (x-\mu)^2 f(x) dx$$

Figure 4.4 illustrates the distribution spread conveyed by the variance measure.

# Basic Mathematical Calculations

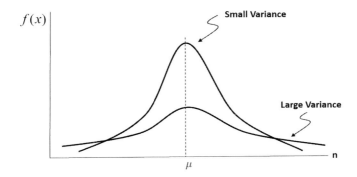

**FIGURE 4.4** Graphical illustration of variance.

## Binomial Distribution

$$f(x) = {}^nc_x p^x (1-p)^{n-x}$$

## Poisson Distribution

$$f(x) = \frac{(np)^x e^{-np}}{x!}$$

## Mean of a Binomial Distribution

$$\mu = np$$

## Variance

$$\sigma^2 = npq$$

where $q = 1 - p$ and is the probability of obtaining $x$ failures in the $n$ trials.

## Normal Distribution

$$f(x) = \frac{1}{\sigma\sqrt{2\pi}} e^{\frac{-(x-\mu)^2}{2\sigma^2}}$$

## Cumulative Distribution Function

$$F(x) = P(X \leq x) = \frac{1}{\sigma\sqrt{2\pi}} \int_{-\infty}^{x} e^{\frac{-(x-\mu)^2}{2\sigma^2}} dx$$

## Population Mean

$$\mu_{\bar{x}} = \mu$$

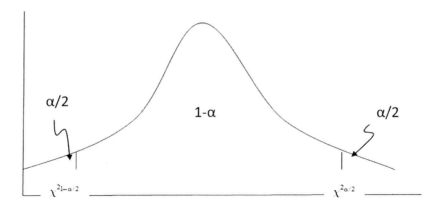

**FIGURE 4.5** Chi-squared distribution.

## Standard Error of the Mean

$$\sigma_{\bar{x}} = \frac{\sigma}{\sqrt{n}}$$

## t-Distribution

$$\bar{x} - t_{\alpha/2}\left(\frac{s}{\sqrt{n}}\right) \le \mu \le \bar{x} + t_{\alpha/2}\left(\frac{s}{\sqrt{n}}\right)$$

where
$\bar{x}$ = sample mean
$\mu$ = population mean
$s$ = sample standard deviation

## Chi-Squared Distribution

The chi-squared distribution is shown graphically in Figure 4.5.

$$\frac{(n-1)s^2}{\chi^{2}_{\alpha/2}} \le \sigma^2 \le \frac{(n-1)s^2}{\chi^{2}_{1-\alpha/2}}$$

## DEFINITION OF SET AND NOTATION

A set is a collection of objects called elements. In mathematics, we write a set by putting its elements between the curly braces { }.

Set *A* containing numbers 3, 4, and 5 is written as

$$A = \{3,4,5\}$$

Basic Mathematical Calculations

a. Empty set
A set with no elements is called an empty set and it denoted by

$$\{\ \} = \Phi$$

b. Subset
Sometimes every element of one set also belongs to another set:

$$A = \{3,4,5\} \text{ and } B = \{1,2,3,4,5,6,7,\}$$

Set A is a subset of set B because every elements of set A is also an element of set B, and it is written as

$$A \subseteq B$$

c. Set equality
Sets A and B are equal if and only if they have exactly the same elements, and the equality is written as

$$A = B$$

d. Set union
The union of set A and set B is the set of all elements that belong to either A or B or both, and it is written as

$$A \cup b = \{x | x \in A \text{ or } x \in B \text{ or both}\}$$

## SET TERMS AND SYMBOLS

{ }: set braces
$\in$: is an element of
$\notin$: is not an element of
$\subseteq$: is a subset of
$\not\subset$: is not a subset of
A': complement of set A
$\cap$: set intersection
$\cup$: set union

## VENN DIAGRAMS

Venn diagrams are used to visually illustrate relationships between sets. Examples are shown in Figure 4.6.

These Venn diagrams illustrate the following statements:

a. Set A is a subset of set B $(A \subset B)$.
b. Set B' is the complement of B.
c. Two sets A and B with their intersection $A \cap B$.
d. Two sets A and B with their union $A \cup B$.

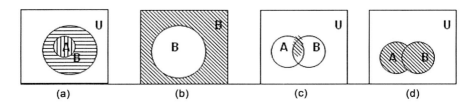

**FIGURE 4.6** Venn diagram examples.

## OPERATIONS ON SETS

If $A$, $B$, and $C$ are arbitrary subsets of universal set $U$, then the following rules govern the operations on sets:

1. Commutative law for union

$$A \cup B = B \cup A$$

2. Commutative law for intersection

$$A \cap B = B \cap A$$

3. Associative law for union

$$A \cup (B \cup C) = (A \cup B) \cup C$$

4. Associative law for intersection

$$A \cap (B \cap C) = (A \cap B) \cap C$$

5. Distributive law for union

$$A \cup (B \cap C) = (A \cup B) \cap (A \cup C)$$

6. Distributive law for intersection

$$A \cap (B \cap C) = (A \cap B) \cup (A \cap C)$$

## DE MORGAN'S LAWS

$$(A \cup B)' = A' \cap B' \tag{4.1}$$

$$(A \cap B)' = A' \cup B' \tag{4.2}$$

The complement of the union of two sets is equal to the intersection of their complements (Eq. 4.1). The complement of the intersection of two sets is equal to the union of their complements (Eq. 4.2).

Counting the Elements Is a Set

The number of the elements in a finite set is determined by simply counting the elements in the set.

If A and B are disjoint sets, then

$$n(A \cup B) = n(A) + n(B)$$

In general, A and B need not to be disjoint, so

$$n(A \cup B) = n(A) + n(B) - n(A \cap B)$$

where
$n$ = number of the elements in a set

## PROBABILITY TERMINOLOGY

A number of specialized terms are used in the study of probability.

**Experiment.** An activity or occurrence with an observable result
**Outcome.** The result of the experiment
**Sample point.** An outcome of an experiment
**Event.** A set of outcomes (a subset of the sample space) to which a probability is assigned

## BASIC PROBABILITY PRINCIPLES

Consider a random sampling process in which all the outcomes solely depend on chance, that is, each outcome is equally likely to happen. If $S$ is a uniform sample space and the collection of desired outcomes is $E$, the probability of the desired outcomes is

$$P(E) = \frac{n(E)}{n(S)}$$

where
$n(E)$ = number of favorable outcomes in $E$
$n(S)$ = number of possible outcomes in $S$

Since $E$ is a subset of $S$,

$$0 \leq n(E) \leq n(S),$$

the probability of the desired outcome is

$$0 \leq P(E) \leq 1$$

## RANDOM VARIABLE

A random variable is a rule that assigns a number to each outcome of a chance experiment.

Example:

1. A coin is tossed six times. The random variable $X$ is the number of tails that are noted. $X$ can only take the values 1, 2, ..., 6, so $X$ is a discrete random variable.
2. A light bulb is burned until it burns out. The random variable $Y$ is its lifetime in hours. $Y$ can take any positive real value, so $Y$ is a continuous random variable.

## MEAN VALUE $\hat{x}$ OR EXPECTED VALUE $\mu$

The mean value or expected value of a random variable indicates its average or central value. It is a useful summary value of the variable's distribution.

1. If random variable $X$ is a discrete mean value,

$$\hat{x} = x_1 p_1 + x_2 p_2 + \ldots + x_n p_n = \sum_{i=1}^{n} x_1 p_1$$

where
$p_i$ = probability densities

2. If $X$ is a continuous random variable with probability density function $f(x)$, then the expected value of $X$ is

$$\mu = E(X) = \int_{-\infty}^{+\infty} x f(x) dx$$

where
$f(x)$ = probability densities

## SERIES EXPANSIONS

a. Expansions of common functions

$$e = 1 + \frac{1}{1!} + \frac{1}{2!} + \frac{1}{3!} + \cdots$$

$$e^x = 1 + x + \frac{x^2}{2!} + \frac{x^3}{3!} + \cdots$$

$$a^x = 1 + x \ln a + \frac{(x \ln a)^2}{2!} + \frac{(x \ln a)^3}{3!} + \cdots$$

$$e^{-x^2} = 1 - x^2 + \frac{x^4}{2!} - \frac{x^6}{3!} + \frac{x^8}{4!} - \cdots$$

# Basic Mathematical Calculations

$$\ln x = (x-1) - \frac{1}{2}(x-1)^2 + \frac{1}{3}(x-1)^3 - \cdots, \qquad 0 < x \leq 2$$

$$\ln x = \frac{x-1}{x} + \frac{1}{2}\left(\frac{x-1}{x}\right)^2 + \frac{1}{3}\left(\frac{x-1}{x}\right)^3 + \cdots, \qquad x > \frac{1}{2}$$

$$\ln x = 2\left[\frac{x-1}{x+1} + \frac{1}{3}\left(\frac{x-1}{x+1}\right)^3 + \frac{1}{5}\left(\frac{x-1}{x+1}\right)^5 + \cdots\right], \qquad x > 0$$

$$\ln(1+x) = x - \frac{x^2}{2} + \frac{x^3}{3} - \frac{x^4}{4} + \cdots, \qquad |x| \leq 1$$

$$\ln(a+x) = \ln a + 2\left[\frac{x}{2a+x} + \frac{1}{3}\left(\frac{x}{2a+x}\right)^3 + \frac{1}{5}\left(\frac{x}{2a+x}\right)^5 + \cdots\right],$$

$$a > 0, \quad -a < x < +\infty$$

$$\ln\left(\frac{1+x}{1-x}\right) = 2\left(x + \frac{x^3}{3} + \frac{x^5}{5} + \frac{x^7}{7} + \cdots\right), \qquad x^2 < 1$$

$$\ln\left(\frac{1+x}{1-x}\right) = 2\left[\frac{1}{x} + \frac{1}{3}\left(\frac{1}{x}\right)^3 + \frac{1}{5}\left(\frac{1}{x}\right)^5 + \left(\frac{1}{x}\right)^7 + \cdots\right], \qquad x^2 > 1$$

$$\ln\left(\frac{1+x}{x}\right) = 2\left[\frac{1}{2x+1} + \frac{1}{3(2x+1)^3} + \frac{1}{5(2x+1)^5} + \cdots\right], \qquad x > 0$$

$$\sin x = x - \frac{x^3}{3!} + \frac{x^5}{5!} - \frac{x^7}{7!} + \cdots$$

$$\cos x = 1 - \frac{x^2}{2!} + \frac{x^4}{4!} - \frac{x^6}{6!} + \cdots$$

$$\tan x = x + \frac{x^3}{3} + \frac{2x^5}{15} + \frac{17x^7}{315} + \frac{62x^9}{2835} + \cdots, \qquad x^2 < \frac{\pi^2}{4}$$

$$\sin^{-1} x = x + \frac{x^3}{6} + \frac{1}{2} \cdot \frac{3}{4} \cdot \frac{x^3}{5} + \frac{1}{2} \cdot \frac{3}{4} \cdot \frac{5}{6} \cdot \frac{x^7}{7} + \cdots, \qquad x^2 < 1$$

$$\tan^{-1} x = x - \frac{1}{3}x^3 + \frac{1}{5}x^5 - \frac{1}{7}x^7 + \cdots, \qquad x^2 < 1$$

$$\tan^{-1} x = \frac{\pi}{2} - \frac{1}{x} + \frac{1}{3x^3} - \frac{1}{5x^5} + \cdots, \qquad x^2 > 1$$

$$\sinh x = x + \frac{x^3}{3!} + \frac{x^5}{5!} + \frac{x^7}{7!} + \cdots$$

$$\cosh x = 1 + \frac{x^2}{2!} + \frac{x^4}{4!} + \frac{x^6}{6!} + \cdots$$

$$\tanh x = x - \frac{x^3}{3} + \frac{2x^5}{15} - \frac{17x^7}{315} + \cdots$$

$$\sinh^{-1} x = x - \frac{1}{2} \cdot \frac{x^3}{3} + \frac{1 \cdot 3}{2 \cdot 4} \cdot \frac{x^5}{5} - \frac{1 \cdot 3 \cdot 5}{2 \cdot 4 \cdot 6} \cdot \frac{x^7}{7} + \cdots, \qquad x^2 < 1$$

$$\sinh^{-1} x = \ln 2x + \frac{1}{2} \cdot \frac{1}{2x^2} - \frac{1 \cdot 3}{2 \cdot 4} \cdot \frac{1}{4x^4} + \frac{1 \cdot 3 \cdot 5}{2 \cdot 4 \cdot 6} \cdot \frac{1}{6x^6} - \cdots, \qquad x > 1$$

$$\cosh^{-1} x = \ln 2x - \frac{1}{2} \cdot \frac{1}{2x^2} - \frac{1 \cdot 3}{2 \cdot 4} \cdot \frac{1}{4x^4} - \frac{1 \cdot 3 \cdot 5}{2 \cdot 4 \cdot 6} \cdot \frac{1}{6x^6} - \cdots$$

$$\tanh^{-1} x = x + \frac{x^3}{3} + \frac{x^5}{5} + \frac{x^7}{7} + \cdots, \qquad x^2 < 1$$

b. Binomial theorem

$$(a+x)^n = a^n + na^{n-1}x + \frac{n(n-1)}{2!}a^{n-2}x^2 + \frac{n(n-1)(n-2)}{3!}a^{n-3}x^3 + \cdots, \qquad x^2 < a^2$$

c. Taylor series expansion

A function $f(x)$ may be expanded about $x = a$ if the function is continuous, and its derivatives exist and are finite at $x = a$.

$$f(x) = f(a) + f'(a)\frac{(x-a)}{1!} + f''(a)\frac{(x-a)^2}{2!} + f'''(a)\frac{(x-a)^3}{3!} + \cdots$$

$$+ f^{n-1}(a)\frac{(x-a)^{n-1}}{(n-1)!} + R_n$$

d. Maclaurin series expansion

The Maclaurin series expansion is a special case of the Taylor series expansion for $a = 0$.

$$f(x) = f(0) + f'(0)\frac{x}{1!} + f''(0)\frac{x^2}{2!} + f'''(0)\frac{x^3}{3!} + \cdots + f^{(n-1)}(0)\frac{x^{n-1}}{(n-1)!} + R_n$$

# Basic Mathematical Calculations

e. Arithmetic progression
   The sum to $n$ terms of the arithmetic progression

$$S = a + (a+d) + (a+2d) + \cdots + [a+(n-1)d]$$

is (in terms of the last number $l$)

$$S = \frac{n}{2}(a+l)$$

where

$$l = a + (n-1)d$$

f. Geometric progression
   The sum of the geometric progression to $n$ terms is

$$S = a + ar + ar^2 + \cdots + ar^{n-1} = a\left(\frac{1-r^n}{1-r}\right)$$

g. Sterling's formula for factorials

$$n! \approx \sqrt{2\pi}\, n^{n+1/2} e^{-n}$$

# MATHEMATICAL SIGNS AND SYMBOLS

$\pm$ ($\mp$): plus or minus (minus or plus)
$::$ divided by, ratio sign
$:::$ proportional sign
$<$: less than
$\not<$: not less than
$>$: greater than
$\not>$: not greater than
$\cong$: approximately equals, congruent
$\sim$: similar to
$\equiv$: equivalent to
$\neq$: not equal to
$\doteq$: approaches, approximately equal to
$\propto$: varies as
$\infty$: infinity
$\therefore$: therefore
$\sqrt{\phantom{x}}$: square root
$\sqrt[3]{\phantom{x}}$: cube root
$\sqrt[n]{\phantom{x}}$: $n$th root
$\angle$: angle
$\perp$: perpendicular to

$\|$: parallel to
$|x|$: numerical value of $x$
log or $\log_{10}$: common logarithm or Briggsian logarithm
$\log_e$ or ln: natural logarithm or hyperbolic logarithm or Napierian logarithm
$e$: base (2.718) of natural system of logarithms
$a°$: an angle "$a$" in degrees
$a'$: a prime, an angle "$a$" in minutes
$a''$: a double prime, an angle "$a$" in seconds, a second
sin: sine
cos: cosine
tan: tangent
ctn or cot: cotangent
sec: secant
csc: cosecant
vers: versed sine
covers: coversed sine
exsec: exsecant
$\sin^{-1}$: anti-sine or angle whose sine is
sinh: hyperbolic sine
cosh: hyperbolic cosine
tanh: hyperbolic tangent
$\sinh^{-1}$: anti-hyperbolic sine or angle whose hyperbolic sine is
$f(x)$ or $\phi(x)$: function of $x$
$\Delta x$: increment of $x$
$\sum$: summation of
$dx$: differential of $x$
$dy/dx$ or $y'$: derivative of $y$ with respect to $x$
$d^2y/dx^2$ or $y''$: second derivative of $y$ with respect to $x$
$d^n y/dx^n$: $n$th derivative of $y$ with respect to $x$
$\partial y/\partial x$: partial derivative of $y$ with respect to $x$
$\partial^n y/\partial x^n$: $n$th partial derivative of $y$ with respect to $x$
$\dfrac{\partial^n y}{\partial x \, \partial y}$: $n$th partial derivative with respect to $x$ and $y$
$\int$: integral of
$\int_a^b$: integral between the limits $a$ and $b$
$\dot{y}$: first derivative of $y$ with respect to time
$\ddot{y}$: second derivative of $y$ with respect to time
$\Delta$ or $\nabla^2$: the "Laplacian"
$\delta$: sign of a variation
$\xi$: sign of integration around a closed path

# GREEK ALPHABET

| Name | Letter | Capital | Pronunciation (as in) | English |
|---|---|---|---|---|
| Alpha | α | A | al-fah (hat) | a, A |
| Beta | β | B | bay-tah (ball) | b, B |
| Gamma | γ | Γ | gam-ah (gift) | g, G |
| Delta | δ | Δ | del-tah (den) | d, D |
| Epsilon | ε | E | ep-si-lon (met) | e, E |
| Zeta | ζ | Z | zay-tah (zoo) | z, Z |
| Eta | η | H | ay-tay, ay-tah (they) | e, E |
| Theta | θ | Θ | thay-tah (thing) | Th |
| Iota | ι | I | eye-o-tah (kit) | i, I |
| Kappa | κ | K | cap-ah (kitchen) | k, K |
| Lambda | λ | Λ | lamb-dah (lamb) | l, L |
| Mu | μ | M | mew (mother) | m, M |
| Nu | ν | N | new (nice) | n, N |
| Xi | ξ | Ξ | zzEee, zee-eye (taxi) | x, X |
| Omicron | o | O | om-ah-cron (pot) | o, O |
| Pi | π | Π | pie (pie) | p, P |
| Rho | ρ | P | row (row) | r, R |
| Sigma | σ, ς | Σ | sig-ma (sigma) | s, S |
| Tau | τ | T | tawh (tau) | t, T |
| Upsilon | υ | Υ | oop-si-lon (put) | U, U |
| Phi | φ | Φ | figh, fie, fah-ee (phone) | Ph |
| Chi | χ | X | kigh (kah-i) | Ch |
| Psi | ψ | Ψ | sigh (sigh) | Ps |
| Omega | ω | Ω | o-may-gah (bone) | O |

# ALGEBRA

### Laws of Algebraic Operations

a. Commutative law: $a + b = b + a, \quad ab = ba$
b. Associative law: $a + (b + c) = (a + b) + c, \quad a(bc) = (ab)c$
c. Distributive law: $c(a + b) = ca + cb$

### Special Products and Factors

$$(x + y)^2 = x^2 + 2xy + y^2$$

$$(x - y)^2 = x^2 - 2xy + y^2$$

$$(x + y)^3 = x^3 + 3x^2 y + 3xy^2 + y^3$$

$$(x-y)^3 = x^3 - 3x^2y + 3xy^2 - y^3$$

$$(x+y)^4 = x^4 + 4x^3y + 6x^2y^2 + 4xy^3 + y^4$$

$$(x-y)^4 = x^4 - 4x^3y + 6x^2y^2 - 4xy^3 + y^4$$

$$(x+y)^5 = x^5 + 5x^4y + 10x^3y^2 + 10x^2y^3 + 5xy^4 + y^5$$

$$(x-y)^5 = x^5 - 5x^4y + 10x^3y^2 - 10x^2y^3 + 5xy^4 - y^5$$

$$(x+y)^6 = x^6 + 6x^5y + 15x^4y^2 + 20x^3y^3 + 15x^2y^4 + 6xy^5 + y^6$$

$$(x-y)^6 = x^6 - 6x^5y + 15x^4y^2 - 20x^3y^3 + 15x^2y^4 - 6xy^5 + y^6$$

The results above are special cases of the binomial formula.

$$x^2 - y^2 = (x-y)(x+y)$$

$$x^3 - y^3 = (x-y)(x^2 + xy + y^2)$$

$$x^3 + y^3 = (x+y)(x^2 - xy + y^2)$$

$$x^4 - y^4 = (x-y)(x+y)(x^2 + y^2)$$

$$x^5 - y^5 = (x-y)(x^4 + x^3y + x^2y^2 + xy^3 + y^4)$$

$$x^5 + y^5 = (x+y)(x^4 - x^3y + x^2y^2 - xy^3 + y^4)$$

$$x^6 - y^6 = (x-y)(x+y)(x^2 + xy + y^2)(x^2 - xy + y^2)$$

$$x^4 + x^2y^2 + y^4 = (x^2 + xy + y^2)(x^2 - xy + y^2)$$

$$x^4 + 4y^4 = (x^2 + 2xy + 2y^2)(x^2 - 2xy + 2y^2)$$

Some generalization of the above are given by the following results where $n$ is a positive integer:

$$x^{2n+1} - y^{2n+1} = (x-y)(x^{2n} + x^{2n-1}y + x^{2n-2}y^2 + \cdots + y^{2n})$$

$$= (x-y)\left(x^2 - 2xy\cos\frac{2\pi}{2n+1} + y^2\right)\left(x^2 - 2xy\cos\frac{4\pi}{2n+1} + y^2\right)$$

$$\cdots \left(x^2 - 2xy\cos\frac{2n\pi}{2n+1} + y^2\right)$$

$$x^{2n+1} + y^{2n+1} = (x+y)\left(x^{2n} - x^{2n-1}y + x^{2n-2}y^2 - \cdots + y^{2n}\right)$$

$$= (x+y)\left(x^2 + 2xy\cos\frac{2\pi}{2n+1} + y^2\right)\left(x^2 + 2xy\cos\frac{4\pi}{2n+1} + y^2\right)$$

$$\cdots \left(x^2 + 2xy\cos\frac{2n\pi}{2n+1} + y^2\right)$$

$$x^{2n} - y^{2n} = (x-y)(x+y)\left(x^{n-1} + x^{n-2}y + x^{n-3}y^2 + \cdots\right)\left(x^{n-1} - x^{n-2}y + x^{n-3}y^2 - \cdots\right)$$

$$= (x-y)(x+y)\left(x^2 - 2xy\cos\frac{\pi}{n} + y^2\right)\left(x^2 - 2xy\cos\frac{2\pi}{n} + y^2\right)$$

$$\cdots \left(x^2 - 2xy\cos\frac{(n-1)\pi}{n} + y^2\right)$$

$$x^{2n} + y^{2n} = \left(x^2 + 2xy\cos\frac{\pi}{2n} + y^2\right)\left(x^2 + 2xy\cos\frac{3\pi}{2n} + y^2\right)$$

$$\cdots \left(x^2 + 2xy\cos\frac{(2n-1)\pi}{2n} + y^2\right)$$

## Powers and Roots

$$a^x \times a^y = a^{(x+y)} \qquad a^0 = 1\,[\text{if } a \neq 0] \qquad (ab^x) = a^x b^x$$

$$\frac{a^x}{a^y} = a^{(x-y)} \qquad a^{-x} = \frac{1}{a^x} \qquad \left(\frac{a}{b}\right)^x = \frac{a^x}{b^x}$$

$$(a^x)^y = a^{xy} \qquad a^{\frac{1}{x}} = \sqrt[x]{a} \qquad \sqrt[x]{ab} = \sqrt[x]{a}\,\sqrt[x]{b}$$

$$\sqrt[x]{\sqrt[y]{a}} = \sqrt[xy]{a} \qquad a^{\frac{x}{y}} = \sqrt[y]{a^x} \qquad \sqrt[x]{\frac{a}{b}} = \frac{\sqrt[x]{a}}{\sqrt[x]{b}}$$

## Proportion

If $\dfrac{a}{b} = \dfrac{c}{d}$, then $\dfrac{a+b}{b} = \dfrac{c+d}{d}$

$$\frac{a-b}{b} = \frac{c-d}{d} \qquad \frac{a-b}{a+b} = \frac{c-d}{c+d}$$

Sum of arithmetic progression to $n$ terms[1]

$$a + (a+d) + (a+2d) + \ldots + (a+(n-1)d)$$

$$= na + \frac{1}{2}n(n-1)d = \frac{n}{2}(a+l),$$

the last term in series = $l = a + (n-1)d$

Sum of geometric progression to $n$ terms

$$s_n = a + ar + ar^2 + \ldots + ar^{n-1} = \frac{a(1-r^n)}{1-r}$$

$$\lim_{n \to \infty} 8_n = a!(1-r) \qquad (-1 < r < 1)$$

### ARITHMETIC MEAN OF N QUANTITIES A

$$A = \frac{a_1 + a_2 + \cdots + a_n}{n}$$

### GEOMETRIC MEAN OF N QUANTITIES G

$$G = (a_1 a_2 \ldots a_n)^{1/n}$$

$$(a_k > 0, \ k = 1, 2, \ldots, n)$$

### HARMONIC MEAN OF N QUANTITIES H

$$\frac{1}{H} = \frac{1}{n}\left(\frac{1}{a_1} + \frac{1}{a_2} + \ldots + \frac{1}{a_n}\right)$$

$$(a_k > 0, \ k = 1, 2, \ldots, n)$$

### GENERALIZED MEAN

$$M(t) = \left(\frac{1}{n}\sum_{k=1}^{n} a_k^t\right)^{1/t}$$

$$M(t) = 0 (t < 0, \ \text{some } a_k \ \text{zero})$$

$$\lim_{t \to \infty} M(t) = \max. \qquad (a_1, a_2, \ldots, a_n) = \max. a$$

$$\lim_{t \to -\infty} M(t) = \min. \qquad (a_1, a_2, \ldots, a_n) = \min. a$$

$$\lim_{t \to 0} M(t) = G$$

$$M(1) = A$$

$$M(-1) = H$$

# Basic Mathematical Calculations

## SOLUTION OF QUADRATIC EQUATIONS

Given $az^2 + bz + c = 0$

$$z_{1,2} = -\left(\frac{b}{2a}\right) \pm \frac{1}{2a} q^{\frac{1}{2}}, \quad q = b^2 - 4ac,$$

$$z_1 + z_2 = -b/a, \quad z_1 z_2 = c/a$$

If $q > 0$, two real roots.
If $q = 0$, two equal roots.
If $q < 0$, pair of complex conjugate roots.

## SOLUTION OF CUBIC EQUATIONS

Given $z^3 + a_2 z^2 + a_1 z + a_0 = 0$, let

$$q = \frac{1}{3} a_1 - \frac{1}{9} a_2^2$$

$$r = \frac{1}{6}(a_1 a_2 - 3a_0) - \frac{1}{27} a_2^3$$

If $q^3 + r^2 > 0$, one real root and a pair of complex conjugate roots.
If $q^3 + r^2 = 0$, all roots real and at least two are equal.
If $q^3 + r^2 < 0$, all roots real (irreducible case).
Let

$$s_1 = \left[r + \left(q^3 + r^2\right)^{\frac{1}{2}}\right]^{\frac{1}{2}}$$

$$s_2 = \left[r - \left(q^3 + r^2\right)^{\frac{1}{2}}\right]^{\frac{1}{2}}$$

then

$$z_1 = (s_1 + s_2) - \frac{a_2}{3}$$

$$z_2 = -\frac{1}{2}(s_1 + s_2) - \frac{a_2}{3} + \frac{i\sqrt{3}}{2}(s_1 - s_2)$$

$$z_3 = -\frac{1}{2}(s_1 + s_2) - \frac{a_2}{3} - \frac{i\sqrt{3}}{2}(s_1 - s_2)$$

If $z_1, z_2, z_3$ are the roots of the cubic equation,

$$z_1 + z_2 + z_3 = -a_2$$

$$z_1 z_2 + z_1 z_3 + z_2 z_3 = a_1$$

$$z_1 z_2 z_3 = a_0$$

## TRIGONOMETRIC SOLUTION OF THE CUBIC EQUATION

The form $x^3 + ax + b = 0$ with $ab \neq 0$ can always be solved by transforming it to the trigonometric identity

$$4\cos^3\theta - 3\cos\theta - \cos(3\theta) \equiv 0$$

Let $x = m\cos\theta$, then

$$x^3 + ax + b = m^3\cos^3\theta + am\cos\theta + b = 4\cos^3\theta - 3\cos\theta - \cos(3\theta) \equiv 0$$

Hence,

$$\frac{4}{m^3} = -\frac{3}{am} = \frac{-\cos(3\theta)}{b}$$

from which follows that

$$m = 2\sqrt{-\frac{a}{3}}, \quad \cos(3\theta) = \frac{3b}{am}$$

Any solution $\theta_1$ which satisfies $\cos(3\theta) = \dfrac{3b}{am}$ will also have the solutions

$$\theta_1 + \frac{2\pi}{3} \quad \text{and} \quad \theta_1 + \frac{4\pi}{3}$$

The roots of the cubic $x^3 + ax + b = 0$ are

$$2\sqrt{-\frac{a}{3}}\cos\theta_1$$

$$2\sqrt{-\frac{a}{3}}\cos\left(\theta_1 + \frac{2\pi}{3}\right)$$

$$2\sqrt{-\frac{a}{3}}\cos\left(\theta_1 + \frac{4\pi}{3}\right)$$

## Solution of Quadratic Equations

Given $z^4 + a_3 z^3 + a_2 z^2 + a_1 z + a_0 = 0$, find the real root $u_1$ of the cubic equation

$$u^3 - a_2 u^2 + (a_1 a_3 - 4a_0)u - (a_1^2 + a_0 a_3^2 - 4a_0 a_2) = 0$$

and determine the four roots of the quartic as solutions of the two quadratic equations

$$v^2 + \left[\frac{a_3}{2} \mp \left(\frac{a_3^2}{4} + u_1 - a_2\right)^{\frac{1}{2}}\right]v + \frac{u_1}{2} \mp \left[\left(\frac{u_1}{2}\right)^2 - a_0\right]^{\frac{1}{2}} = 0$$

If all roots of the cubic equation are real, use the value of $u_1$ which gives real coefficients in the quadratic equation and select signs so that if

$$z^4 + a_3 z^3 + a_2 z^3 + a_1 z + a_0 = (z^2 + p_1 z + q_1)(z^2 + p_2 z + q_2)$$

then

$$p_1 + p_2 = a_3,\; p_1 p_2 + q_1 + q_2 = a_2,\; p_1 q_2 + p_2 q_1 = a_1,\; q_1 q_2 = a_0$$

If $z_1, z_2, z_3, z_4$ are the roots,

$$\sum z_t = -a_3,\; \sum z_t z_j z_k = -a_1$$

$$\sum z_t z_j = a_2,\; z_1 z_2 z_3 z_4 = a_0$$

## Partial Fractions

This section applies only to rational algebraic fractions with the numerator of lower degree than the denominator. Improper fractions can be reduced to proper fractions by long division.

Every fraction may be expressed as the sum of component fractions whose denominators are factors of the denominator of the original fraction.

Let $N(x)$ = numerator, a polynomial of the form

$$N(x) = n_0 + n_1 x + n_2 x^2 + \cdots + n_l x^l$$

### Non-repeated Linear Factors

$$\frac{N(x)}{(x-a)G(x)} = \frac{A}{x-a} + \frac{F(x)}{G(x)}$$

$$A = \left[\frac{N(x)}{G(x)}\right]_{x=a}$$

$F(x)$ is determined by methods discussed in the sections that follow.

## REPEATED LINEAR FACTORS

$$\frac{N(x)}{x^m G(x)} = \frac{A_0}{x^m} + \frac{A_1}{x^{m-1}} + \ldots + \frac{A_{m-1}}{x} + \frac{F(x)}{G(x)}$$

$$N(x) = n_o + n_1 x + n_2 x^2 + n_3 x^3 + \ldots$$

$$F(x) = f_0 + f_1 x + f_2 x^2 + \ldots$$

$$G(x) = g_0 + g_1 x + g_2 x^2 + \cdots$$

$$A_0 = \frac{n_0}{g_0}, \quad A_1 = \frac{n_1 - A_0 g_1}{g_0}$$

$$A_2 = \frac{n_2 - A_0 g_2 - A_1 g_1}{g_0}$$

## GENERAL TERMS

$$A_0 = \frac{n_0}{g_0}, \quad A_k = \frac{1}{g_0}\left[n_k - \sum_{t=0}^{k-1} A_t g_{k-t}\right] k \geq 1$$

$$m^* = 1 \begin{cases} f_0 = n_1 - A_0 g_1 \\ f_1 = n_2 - A_0 g_2 \\ f_1 = n_{j+1} - A_0 g_{t+1} \end{cases}$$

$$m = 2 \begin{cases} f_0 = n_2 - A_0 g_2 - A_1 g_1 \\ f_1 = n_3 - A_0 g_3 - A_1 g_2 \\ f_1 = n_{j+2} - [A_0 g_{1+2} + A_1 g_1 + 1] \end{cases}$$

$$m = 3 \begin{cases} f_0 = n_3 - A_0 g_3 - A_1 g_2 - A_2 g_1 \\ f_1 = n_3 - A_0 g_4 - A_1 g_3 - A_2 g_2 \\ f_1 = n_{j+3} - [A_0 g_{j+3} + A_1 g_{j+2} + A_2 g_{j+1}] \end{cases}$$

$$\text{any } m: f_1 = n_{m+1} - \sum_{i=0}^{m-1} A_1 g_{m+j-1}$$

$$\frac{N(x)}{(x-a)^m G(x)} = \frac{A_0}{(x-a)^m} + \frac{A_1}{(x-a)^{m-1}} + \cdots + \frac{A_{m-1}}{(x-a)} + \frac{F(x)}{G(x)}$$

Change to form $\dfrac{N'(y)}{y^m G'(y)}$ by substitution of $x = y + a$. Resolve into partial fractions in terms of $y$ as described above. Then express in terms of $x$ by substitution $y = x - a$.

## REPEATED LINEAR FACTORS

Alternative method of determining coefficients:

$$\frac{N(x)}{(x-a)^m G(x)} = \frac{A_0}{(x-a)^m} + \cdots + \frac{A_k}{(x-a)^{m-k}} + \cdots + \frac{A_{m-1}}{x-a} + \frac{F(x)}{G(x)}$$

$$A_k = \frac{1}{k!} \left\{ D_x^k \left[ \frac{N(x)}{G(x)} \right] \right\}_{x-G}$$

where $D_x^k$ is the differentiating operator, and the derivative of zero order is defined as

$$D_x^0 u = u$$

## FACTORS OF HIGHER DEGREE

Factors of higher degree have the corresponding numerators indicated.

$$\frac{N(x)}{(x^2 + h_1 x + h_0) G(x)} = \frac{a_1 x + a_0}{x^2 + h_1 x + h_0} + \frac{F(x)}{G(x)}$$

$$\frac{N(x)}{(x^2 + h_1 x + h_0)^2 G(x)} = \frac{a_1 x + a_0}{(x^2 + h_1 x + h_0)^2} + \frac{b_1 x + b_0}{(x^2 + h_1 x + h_0)} + \frac{F(x)}{G(x)}$$

$$\frac{N(x)}{(x^3 + h_2 x^2 + h_1 x + h_0) G(x)} = \frac{a_2 x^2 + a_1 x + a_0}{x^3 + h_2 x^2 + h_1 x + h_0} + \frac{F(x)}{G(x)}, \text{ etc.}$$

Problems of this type are determined first by solving for the coefficients due to linear factors as shown above, and then determining the remaining coefficients by the general methods given below.

## GEOMETRY

Mensuration formulas are used for measuring angles and distances in geometry. Examples are presented below.

### TRIANGLES

Let $K$ = area, $r$ = radius of the inscribed circle, and $R$ = radius of circumscribed circle.

## Right Triangle

$$A + B = C = 90°$$

$$c^2 = a^2 + b^2 \quad (\text{Pythagorean relations})$$

$$a = \sqrt{(c+b)(c-b)}$$

$$K = \frac{1}{2}ab$$

$$r = \frac{ab}{a+b+c}, \quad R = \frac{1}{2}c$$

$$h = \frac{ab}{c}, \quad m = \frac{b^2}{c}, \quad n = \frac{a^2}{c}$$

## Equilateral Triangle

$$A = B = C = 60°$$

$$K = \frac{1}{4}a^2\sqrt{3}$$

$$r = \frac{1}{6}a\sqrt{3}, \quad R = \frac{1}{3}a\sqrt{3}$$

$$h = \frac{1}{2}a\sqrt{3}$$

## General Triangle

Let $s = \frac{1}{2}(a+b+c)$, $h_c$ = length of altitude on side $c$, $t_c$ = length of bisector of angle $C$, and $m_c$ = length of median to side $c$.

$$A + B + C = 180°$$

$$c^2 = a^2 + b^2 - 2ab \cos C \quad (\text{law of cosines})$$

$$K = \frac{1}{2}h_c c = \frac{1}{2}ab \sin C$$

$$= \frac{c^2 \sin A \sin B}{2 \sin C}$$

$$= rs = \frac{abc}{4R}$$

# Basic Mathematical Calculations

$$= \sqrt{s(s-a)(s-b)(s-c)} \quad (\text{Heron's formula})$$

$$r = c\sin\frac{A}{2}\sin\frac{B}{2}\sec\frac{C}{2} = \frac{ab\sin C}{2s} = (s-c)\tan\frac{C}{2}$$

$$= \sqrt{\frac{(s-a)(s-b)(s-c)}{s}} = \frac{K}{s} = 4R\sin\frac{A}{2}\sin\frac{B}{2}\sin\frac{C}{2}$$

$$R = \frac{c}{2\sin C} = \frac{abc}{4\sqrt{s(s-a)(s-b)(s-c)}} = \frac{abc}{4K}$$

$$h_c = a\sin B = b\sin A = \frac{2K}{c}$$

$$t_c = \frac{2ab}{a+b}\cos\frac{C}{2} = \sqrt{ab\left\{1 - \frac{c^2}{(a+b)^2}\right\}}$$

$$m_c = \sqrt{\frac{a^2}{2} + \frac{b^2}{2} - \frac{c^2}{4}}$$

## Menelaus's Theorem

A necessary and sufficient condition for points $D$, $E$, $F$ on the respective side lines $BC$, $CA$, $AB$ of a triangle $ABC$ to be collinear is that

$$BD \cdot CE \cdot AF = -DC \cdot EA \cdot FB$$

where all segments in the formula are directed segments.

## Ceva's Theorem

A necessary and sufficient condition for $AD$, $BE$, $CF$, where $D$, $E$, $F$ are points on the respective side lines $BC$, $CA$, $AB$ of a triangle $ABC$, to be concurrent is that

$$BD \cdot CE \cdot AF = +DC \cdot EA \cdot FB$$

where all segments in the formula are directed segments.

## QUADRILATERALS

Let $K$ = area, $p$ and $q$ are diagonals.

### RECTANGLE

$$A = B = C = D = 90°$$

$$K = ab, \quad p = \sqrt{a^2 + b^2}$$

## Parallelogram

$$A = C, \quad B = D, \quad A + B = 180°$$

$$K = bh = ab\sin A = ab\sin B$$

$$h = a\sin A = a\sin B$$

$$p = \sqrt{a^2 + b^2 - 2ab\cos A}$$

$$q = \sqrt{a^2 + b^2 - 2ab\cos B} = \sqrt{a^2 + b^2 + 2ab\cos A}$$

## Rhombus

$$p^2 + q^2 = 4a^2$$

$$K = \frac{1}{2}pq$$

## Trapezoid

$$m = \frac{1}{2}(a+b)$$

$$K = \frac{1}{2}(a+b)h = mh$$

## General Quadrilateral

Let $s = \frac{1}{2}(a+b+c+d)$.

$$K = \frac{1}{2}pq\sin\theta$$

$$= \frac{1}{4}\left(b^2 + d^2 - a^2 - c^2\right)\tan\theta$$

$$= \frac{1}{4}\sqrt{4p^2q^2 - \left(b^2 + d^2 - a^2 - c^2\right)^2}$$

(Bretschneider's formula)

$$= \sqrt{(s-a)(s-b)(s-c)(s-d) - abcd\cos^2\left(\frac{A+B}{2}\right)}$$

**Theorem:**
The diagonals of a quadrilateral with consecutive sides $a, b, c, d$ are perpendicular if and only if $a^2 + c^2 = b^2 + d^2$.

## REGULAR POLYGON OF N SIDES EACH OF LENGTH B

$$\text{Area} = \frac{1}{4}nb^2 \cot\frac{\pi}{n} = \frac{1}{4}nb^2 \frac{\cos(\pi/n)}{\sin(\pi/n)}$$

Perimeter = $nb$
Circle of radius = $r$

$$\text{Area} = \pi r^2$$

$$\text{Perimeter} = 2\pi r$$

## REGULAR POLYGON OF N SIDES INSCRIBED IN A CIRCLE OF RADIUS R

$$\text{Area} = \frac{1}{2}nr^2 \sin\frac{2\pi}{n} = \frac{1}{2}nr^2 \sin\frac{360°}{n}$$

$$\text{Perimeter} = 2nr\sin\frac{\pi}{n} = 2nr\sin\frac{180°}{n}$$

## REGULAR POLYGON OF N SIDES CIRCUMSCRIBING A CIRCLE OF RADIUS R

$$\text{Area} = nr^2 \tan\frac{\pi}{n} = nr^2 \tan\frac{180°}{n}$$

$$\text{Perimeter} = 2nr\tan\frac{\pi}{n} = 2nr\tan\frac{180°}{n}$$

## CYCLIC QUADRILATERAL

Let $R$ = radius of the circumscribed circle.

$$A + C = B + D = 180°$$

$$K = \sqrt{(s-a)(s-b)(s-c)(s-d)} = \frac{\sqrt{(ac+bd)(ad+bc)(ab+cd)}}{4R}$$

$$p = \sqrt{\frac{(ac+bd)(ab+cd)}{ad+bc}}$$

$$q = \sqrt{\frac{(ac+bd)(ad+bc)}{ab+cd}}$$

$$R = \frac{1}{2}\sqrt{\frac{(ac+bd)(ad+bc)(ab+cd)}{(s-a)(s-b)(s-c)(s-d)}}$$

$$\sin\theta = \frac{2K}{ac+bd}$$

## PROLEMY'S THEOREM

A convex quadrilateral with consecutive sides $a, b, c, d$ and diagonals $p, q$ is cyclic if and only if $ac + bd = pq$.

## CYCLIC-INSCRIPTABLE QUADRILATERAL

Let $r$ = radius of the inscribed circle, $R$ = radius of the circumscribed circle, and $m$ = distance between the centers of the inscribed and circumscribed circles.

$$A + C = B + D = 180°$$

$$a + c = b + d$$

$$K = \sqrt{abcd}$$

$$\frac{1}{(R-m)^2} + \frac{1}{(R+m)^2} = \frac{1}{r^2}$$

$$r = \frac{\sqrt{abcd}}{s}$$

$$R = \frac{1}{2}\sqrt{\frac{(ac+bd)(ad+bc)(ab+cd)}{abcd}}$$

Sector of circle of radius $r$

$$\text{Area} = \frac{1}{2}r^2\theta \quad [\theta \text{ in radians}]$$

$$\text{Arc length } s = r\theta$$

Radius of circle inscribed in a triangle of sides $a, b, c$

$$r = \frac{\sqrt{8(8-a)(8-b)(8-c)}}{8}$$

where

$$s = \frac{1}{2}(a+b+c) = \text{semiperimeter}$$

Radius of circle circumscribing a triangle of sides $a, b, c$

$$R = \frac{abc}{4\sqrt{8(8-a)(8-b)(8-c)}}$$

where

$$s = \frac{1}{2}(a+b+c) = \text{semiperimeter}$$

Segment of circle of radius $r$

Area of shaded part $= \frac{1}{2}r^2(\theta - \sin\theta)$

Ellipse of semi-major axis $a$ and semi-minor axis $b$

$$\text{Area} = \pi ab$$

$$\text{Perimeter} = 4a\int_0^{\pi/2}\sqrt{1 - k^2\sin^2\theta}\,d\theta$$

$$= 2\pi\sqrt{\frac{1}{2}(a^2 + b^2)} \quad \text{[approximately]}$$

where $k = \sqrt{a^2 - b^2}/a$.

Segment of a parabola

$$\text{Area} = \frac{2}{8}ab$$

$$\text{Arc length } ABC = \frac{1}{2}\sqrt{b^2 + 16a^2} + \frac{b^2}{8a}\ln\left(\frac{4a + \sqrt{b^2 + 16a^2}}{b}\right)$$

## PLANAR AREAS BY APPROXIMATION

Divide the planar area $K$ into $n$ strips by equidistant parallel chords of lengths $y_0, y_1, y_2, \ldots, y_n$ (where $y_0$ and/or $y_n$ may be zero), and let $h$ denote the common distance between the chords.

Then, approximately:

**Trapezoidal Rule**

$$K = h\left(\frac{1}{2}y_0 + y_1 + y_2 + \cdots + y_{n-1} + \frac{1}{2}y_n\right)$$

**Durand's Rule**

$$K = h\left(\frac{4}{10}y_0 + \frac{11}{10}y_1 + y_2 + y_3 + \cdots + y_{n-2} + \frac{11}{10}y_{n-1} + \frac{4}{10}y_n\right)$$

**Simpson's Rule ($n$ even)**

$$K = \frac{1}{3}h\left(y_0 + 4y_1 + 2y_2 + 4y_3 + 2y_4 + \cdots + 2y_{n-2} + 4y_{n-1} + y_n\right)$$

**Weddle's Rule ($n = 6$)**

$$K = \frac{3}{10}h(y_0 + 5y_1 + y_2 + 6y_3 + y_4 + 5y_5 + y_6)$$

## SOLIDS BOUNDED BY PLANES

In the following, $S$ = lateral surface, $T$ = total surface, $V$ = volume.

### Cube
Let $a$ = length of each edge.

$$T = 6a^2, \text{ diagonal of face} = a\sqrt{2}$$

$$V = a^3, \text{ diagonal of cube} = a\sqrt{3}$$

### Rectangular Parallelepiped (or Box)
Let $a$, $b$, $c$, be the lengths of its edges.

$$T = 2(ab + bc + ca), \quad V = abc$$

$$\text{Diagonal} = \sqrt{a^2 + b^2 + c^2}$$

### PRISM

$S$ = (perimeter of the base) × (height) + 2(area of the base): total surface area of a right prism
$V$ = (area of the base) × (height)
Truncated triangular prism
$V$ = (area of right section) × $\frac{1}{3}$ (sum of the three lateral edges)

### PYRAMID

$S$ of regular pyramid = $\frac{1}{2}$ (perimeter of base) × (slant height)

$V = \frac{1}{3}$ (area of base) × (altitude)

Frustum of pyramid
Let $B_1$ = area of lower base, $B_2$ = area of upper base, and $h$ = altitude.
$S$ of regular figure = $\frac{1}{2}$ (sum of perimeters of base) × (slant height)

$$V = \frac{1}{3}h(B_1 + B_2) + \sqrt{B_1 B_2}$$

# Basic Mathematical Calculations

## PRISMATOID

A prismatoid is a polyhedron having for bases two polygons in parallel planes and for lateral faces triangles or trapezoids with one side lying in one base, and the opposite vertex or side lying in the other base, of the polyhedron. Let $B_1$ = area of lower base, $M$ = area of midsection, $B_2$ = area of upper base, and $h$ = altitude.

$$V = \frac{1}{6}h(B_1 + 4M + B_2) \text{ (the prismoidal formula)}$$

Note: Since cubes, rectangular parallelepipeds, prisms, pyramids, and frustums of pyramids are all examples of prismatoids, the formula for the volume of a prismatoid subsumes most of the above volume formulae.

## REGULAR POLYHEDRA

Let
  $v$ = number of vertices
  $e$ = number of edges
  $f$ = number of faces
  $\alpha$ = each dihedral angle
  $a$ = length of each edge
  $r$ = radius of the inscribed sphere
  $R$ = radius of the circumscribed sphere
  $A$ = area of each face
  $T$ = total area
  $V$ = volume

$$v - e + f = 2$$

$$T = fA$$

$$V = \frac{1}{3}rfA = \frac{1}{3}rT$$

| Name | Nature of Surface | $T$ | $V$ |
|---|---|---|---|
| Tetrahedron | Four equilateral triangles | $1.73205\ a^2$ | $0.11785\ a^3$ |
| Hexahedron (cube) | Six squares | $6.00000\ a^2$ | $1.00000\ a^3$ |
| Octahedron | Eight equilateral triangles | $3.46410\ a^2$ | $0.47140\ a^3$ |
| Dodecahedron | Twelve regular pentagons | $20.64573\ a^2$ | $7.66312\ a^3$ |
| Icosahedron | Twenty equilateral triangles | $8.66025\ a^2$ | $2.18169\ a^2$ |

| Name | $v$ | $E$ | $f$ | $\alpha$ | $a$ | $r$ |
|---|---|---|---|---|---|---|
| Tetrahedron | 4 | 6 | 4 | 70° 32′ | 1.633R | 0.333R |
| Hexahedron | 8 | 12 | 6 | 90° | 1.155R | 0.577R |
| Octahedron | 6 | 12 | 8 | 190° 28′ | 1.414R | 0.577R |
| Dodecahedron | 20 | 30 | 12 | 116° 34′ | 0.714R | 0.795R |
| Icosahedron | 12 | 30 | 20 | 138° 11′ | 1.051R | 0.795R |

| Name | $A$ | $r$ | $R$ | $V$ |
|---|---|---|---|---|
| Tetrahedron | $\frac{1}{4}a^2\sqrt{3}$ | $\frac{1}{12}a\sqrt{6}$ | $\frac{1}{4}a\sqrt{6}$ | $\frac{1}{12}a^3\sqrt{2}$ |
| Hexahedron (cube) | $a^2$ | $\frac{1}{2}a$ | $\frac{1}{2}a\sqrt{3}$ | $a^3$ |
| Octahedron | $\frac{1}{4}a^2\sqrt{3}$ | $\frac{1}{6}a\sqrt{6}$ | $\frac{1}{2}a\sqrt{2}$ | $\frac{1}{3}a^3\sqrt{2}$ |
| Dodecahedron | $\frac{1}{4}a^2\sqrt{25+10\sqrt{5}}$ | $\frac{1}{20}a\sqrt{250+110\sqrt{5}}$ | $\frac{1}{4}a(\sqrt{15}+\sqrt{3})$ | $\frac{1}{4}a^3(15+7\sqrt{5})$ |
| Icosahedron | $\frac{1}{4}a^2\sqrt{3}$ | $\frac{1}{12}a\sqrt{42+18\sqrt{5}}$ | $\frac{1}{4}a\sqrt{10+2\sqrt{5}}$ | $\frac{5}{12}a^3(3+\sqrt{5})$ |

## SPHERE OF RADIUS $R$

Volume $= \frac{3}{4}\pi r^3$

Surface area $= 4\pi r^2$

## RIGHT CIRCULAR CYLINDER OF RADIUS $R$ AND HEIGHT $H$

Volume $= \pi r^2 h$
Lateral surface area $= 2\pi rh$

## CIRCULAR CYLINDER OF RADIUS $R$ AND SLANT HEIGHT $\ell$

Volume $= \pi r^2 h = \pi r^2 \ell \sin\theta$
Lateral surface area $= p\ell$

## CYLINDER OF CROSS-SECTIONAL AREA $A$ AND SLANT HEIGHT $\ell$

Volume $= Ah = A\ell \sin\theta$
Lateral surface area $= p\ell$

## Basic Mathematical Calculations

### RIGHT CIRCULAR CONE OF RADIUS R AND HEIGHT H

Volume $= \frac{1}{3}\pi r^2 h$

Lateral surface area $= \pi r \sqrt{r^2 + h^2} = \pi r l$

### SPHERICAL CAP OF RADIUS R AND HEIGHT H

Volume (shaded in figure) $= \frac{1}{3}\pi h^2 (3r - h)$

Surface area $= 2\pi r h$

### FRUSTUM OF RIGHT CIRCULAR CONE OF RADII A, B AND HEIGHT H

Volume $= \frac{1}{3}\pi h (a^2 + ab + b^2)$

Lateral surface area $= \pi(a+b)\sqrt{h^2 + (b-a)^2}$
$= \pi(a+b)l$

### ZONE AND SEGMENT OF TWO BASES

$$S = 2\pi R h = \pi D h$$

$$V = \frac{1}{6}\pi h (3a^2 + 3b^2 + h^2)$$

### LUNE

$S = 2R^3 \theta$, $\theta$ in radians

### SPHERICAL SECTOR

$$V = \frac{2}{3}\pi R^2 h = \frac{1}{6}\pi D^2 h$$

### SPHERICAL TRIANGLE AND POLYGON

Let $A$, $B$, $C$ be the angles, in radians, of the triangle and $\theta$ be the sum of angles, in radians, of a spherical polygon on $n$ sides.

$$S = (A + B + C - \pi)R^2$$

$$S = [\theta - (n-2)\pi]R^2$$

## SPHEROIDS

### Ellipsoid

Let $a$, $b$, $c$ be the lengths of the semiaxes.

$$V = \frac{4}{3}\pi abc$$

### Oblate Spheroid

An oblate spheroid is formed by the rotation of an ellipse about its minor axis. Let $a$ and $b$ be the major and minor semiaxes, respectively, and $\epsilon$ the eccentricity, of the revolving ellipse.

$$S = 2\pi a^2 + \pi \frac{b^2}{\epsilon} \log_e \frac{1+\epsilon}{1-\epsilon}$$

$$V = \frac{4}{3}\pi a^2 b$$

### Prolate Spheroid

A prolate spheroid is formed by the rotation of an ellipse about its major axis. Let $a$ and $b$ be the major and minor semiaxes, respectively, and $\epsilon$ the eccentricity, of the revolving ellipse.

$$S = 2\pi b^2 + 2\pi \frac{ab}{\epsilon} \sin^{-1}\epsilon$$

$$V = \frac{3}{4}\pi ab^2$$

### Circular Torus

A circular torus is formed by the rotation of a circle about an axis in the plane of the circle and not cutting the circle. Let $r$ be the radius of the revolving circle and $R$ be the distance of its center from the axis of rotation.

$$S = 4\pi^2 Rr$$

$$V = 2\pi^2 Rr^2$$

## FORMULAS FROM PLANE ANALYTIC GEOMETRY

### DISTANCE $D$ BETWEEN TWO POINTS

$$P_1(x_1, y_1) \quad \text{and} \quad P_2(x_2, y_2)$$

$$d = \sqrt{(x_2 - x_1)^2 + (y_2 - y_1)^2}$$

# Basic Mathematical Calculations

## SLOPE $M$ OF LINE JOINING TWO POINTS

$$P_1(x_1, y_1) \quad \text{and} \quad P_2(x_2, y_2)$$

$$m = \frac{y_2 - y_1}{x_2 - x_1} = \tan \theta$$

## EQUATION OF LINE JOINING TWO POINTS

$$P_1(x_1, y_1) \quad \text{and} \quad P_2(x_2, y_2)$$

$$\frac{y - y_1}{x - x_1} = \frac{y_2 - y_1}{x_2 - x_1} = m \quad \text{or} \quad y - y_1 = m(x - x_1)$$

$$y = mx + b$$

where $b = y_1 - mx_1 = \dfrac{x_2 y_1 - x_1 y_2}{x_2 - x_1}$ is the intercept on the y axis, that is, the y intercept

## EQUATION OF LINE IN TERMS OF X INTERCEPT $A \neq 0$ AND Y INTERCEPT $b \neq 0$

$$\frac{x}{a} + \frac{y}{b} = 1$$

## NORMAL FORM FOR EQUATION OF LINE

$$x \cos \alpha + y \sin \alpha = p$$

where
 $p$ = perpendicular distance from origin O to line
 $\alpha$ = angle of inclination of perpendicular with positive $x$ axis.

## GENERAL EQUATION OF LINE

$$Ax + By + C = 0$$

## DISTANCE FROM POINT $(x_1, y_1)$ TO LINE $Ax + By + C = 0$

$$\frac{Ax_1 + By_1 + C}{\pm\sqrt{A^2 + B^2}}$$

where the sign is chosen so that the distance is nonnegative.

## ANGLE $\psi$ BETWEEN TWO LINES HAVING SLOPES $m_1$ AND $m_2$

$$\tan \psi = \frac{m_2 - m_1}{1 + m_1 m_2}$$

Lines are parallel or coincident if and only if $m_1 = m_2$.
Lines are perpendicular if and only if $m_2 = -1/m_1$.

## AREA OF TRIANGLE WITH VERTICES

At $(x_1, y_1), (x_2, y_2), (x_3, y_3)$

$$\text{Area} = \pm \frac{1}{2} \begin{vmatrix} x_1 & y_1 & 1 \\ x_2 & y_2 & 1 \\ x_3 & y_3 & 1 \end{vmatrix}$$

$$= \pm \frac{1}{2}(x_1 y_2 + y_1 x_3 + y_3 x_2 - y_2 x_3 - y_1 x_2 - x_1 y_3)$$

where the sign is chosen so that the area is nonnegative.
If the area is zero, the points all lie on a line.

## TRANSFORMATION OF COORDINATES INVOLVING PURE TRANSLATION

$$\begin{cases} x = x' + x_0 \\ y = y' + y_0 \end{cases} \text{ or } \begin{cases} x' = x + x_0 \\ y' = y + y_0 \end{cases}$$

where $x, y$ are old coordinates relative to the $xy$ system, $(x', y')$ are new coordinates relative to the $x'y'$ system, and $(x_0, y_0)$ are the coordinates of the new origin $O'$ relative to the old $xy$ coordinate system.

## TRANSFORMATION OF COORDINATES INVOLVING PURE ROTATION

$$\begin{cases} x = x' \cos \alpha - y' \sin \alpha \\ y = x' \sin \alpha + y' \cos \alpha \end{cases} \text{ or } \begin{cases} x' = x \cos \alpha + y \sin \alpha \\ y' = y \cos \alpha - x \sin \alpha \end{cases}$$

where the origins of the old $(xy)$ and new $(x'y')$ coordinate systems are the same, but the $x'$ axis makes an angle $\alpha$ with the positive $x$ axis.

## TRANSFORMATION OF COORDINATES INVOLVING TRANSLATION AND ROTATION

$$\begin{cases} x = x' \cos \alpha - y' \sin \alpha + x_0 \\ y = x' \sin \alpha + y' \cos \alpha + y_0 \end{cases}$$

or

$$\begin{cases} x' = (x - x_0) \cos \alpha + (y - y_0) \sin \alpha \\ y' = (y - y_0) \cos \alpha - (x - x_0) \sin \alpha \end{cases}$$

# Basic Mathematical Calculations

where the new origin $O'$ of $x'y'$ coordinate system has coordinates $(x_0, y_0)$ relative to the old $xy$ coordinate system and the $x'$ axis makes an angle $\alpha$ with the positive $x$ axis.

## Polar Coordinates $(r, \theta)$

A point $P$ can be located by rectangular coordinates $(x, y)$ or polar coordinates $(r, \theta)$. The transformation between these coordinates is

$$\begin{cases} x = r \cos \theta \\ y = r \sin \theta \end{cases} \text{or} \begin{cases} r = \sqrt{x^2 + y^2} \\ \theta = \tan^{-1}(y/x) \end{cases}$$

## Plane Curves

$$\left(x^2 + y^2\right)^2 = ax^2 y$$

$$r = a \sin \theta \cos^2 \theta$$

## Catenary, Hyperbolic Cosine

$$y = \frac{a}{2}\left(e^{x/e} + e^{-x/e}\right) = a \cosh \frac{x}{a}$$

## Cardioid

$$\left(x^2 + y^2 - ax\right)^2 = a^2\left(x^2 + y^2\right)$$

$$r = a(\cos \theta + 1)$$

or

$$r = a(\cos \theta - 1)$$

$$[P'A = AP = a]$$

## Circle

$$x^2 + y^2 = a^2$$

$$r = a$$

## Cassinian Curves

$$x^2 + y^2 = 2ax$$

$$r = 2a\cos\theta$$
$$x^2 + y^2 = ax + by$$
$$r = a\cos\theta + b\sin\theta$$

## Cotangent Curve

$$y = \cot x$$

## Cubical Parabola

$$y = ax^3, \quad a > 0$$
$$r^2 = \frac{1}{a}\sec^2\theta\tan\theta, \quad a > 0$$

## Cosecant Curve

$$y = \csc x$$

## Cosine Curve

$$y = \cos x$$

## Ellipse

$$x^2/a^2 + y^2/b^2 = 1$$
$$\begin{cases} x = a\cos\phi \\ y = b\sin\phi \end{cases}$$

## Gamma Function

$$\Gamma(n) = \int_0^\infty x^{n-1} e^{-x} dx \quad (n > 0)$$
$$\Gamma(n) = \frac{\Gamma(n+1)}{n} \quad (0 > n \neq -1, -2, -3, \ldots)$$

## Hyperbolic Functions

$$\sinh x = \frac{e^x - e^{-x}}{2} \qquad \operatorname{csch} x = \frac{2}{e^x - e^{-x}}$$

$$\cosh x = \frac{e^x - e^{-x}}{2} \qquad \operatorname{csch} x = \frac{2}{e^x - e^{-x}}$$

$$\tanh x = \frac{e^x - e^{-x}}{e^x + e^{-x}} \qquad \coth x = \frac{e^x + e^{-x}}{e^x - e^{-x}}$$

## Inverse Cosine Curve

$$y = \arccos x$$

## Inverse Sine Curve

$$y = \arcsin x$$

## Inverse Tangent Curve

$$y = \arctan x$$

## Logarithmic Curve

$$y = \log_a x$$

## Parabola

$$y = x^2$$

## Cubical Parabola

$$y = x^3$$

## Tangent Curve

$$y = \tan x$$

## Ellipsoid

$$\frac{x^2}{a^2} + \frac{y^2}{b^2} + \frac{z^2}{c^2} = 1$$

## Elliptic Cone

$$\frac{x^2}{a^2} + \frac{y^2}{b^2} - \frac{z^2}{c^2} = 0$$

## Elliptic Cylinder

$$\frac{x^2}{a^2} + \frac{y^2}{b^2} = 1$$

## Hyperboloid of One Sheet

$$\frac{x^2}{a^2} + \frac{y^2}{b^2} - \frac{z^2}{c^2} = 1$$

## Elliptic Paraboloid

$$\frac{x^2}{a^2} + \frac{y^2}{b^2} = cz$$

## Hyperboloid of Two Sheets

$$\frac{z^2}{c^2} - \frac{x^2}{a^2} - \frac{y^2}{b^2} = 1$$

## Hyperbolic Paraboloid

$$\frac{x^2}{a^2} - \frac{y^2}{b^2} = cz$$

## Sphere

$$x^2 + y^2 + z^2 = a^2$$

## Distance d between Two Points

$$P_1(x_1, y_1, z_1) \text{ and } P_2(x_2, y_2, z_2)$$

$$d = \sqrt{(x_2 - x_1)^2 + (y_2 - y_1)^2 + (z_2 - z_1)^2}$$

## Equations of Line Joining $P_1(x_1, y_1, z_1)$ and $P_2(x_2, y_2, z_2)$ in Standard Form

$$\frac{x - x_1}{x_2 - x_1} = \frac{y - y_1}{y_2 - y_1} = \frac{z - z_1}{z_2 - z_1} \quad \text{or}$$

$$\frac{x - x_1}{l} = \frac{y - y_1}{m} = \frac{z - z_1}{n}$$

## Equations of Line Joining $P_1(x_1, y_1, z_1)$ and $P_2(x_2, y_2, z_2)$ in Parametric Form

$$x = x_1 + lt, \quad y = y_1 + mt, \quad z = z_1 + nt$$

## Angle $\phi$ between Two Lines with Direction Cosines $l_1, m_1, n_1$ and $l_2, m_2, n_2$

$$\cos\phi = l_1 l_2 + m_1 m_2 + n_1 n_2$$

## GENERAL EQUATION OF A PLANE

$$Ax + By + Cz + D = 0$$

where $A$, $B$, $C$, $D$ are constants.

## EQUATION OF PLANE PASSING THROUGH POINTS

$$(x_1, y_1, z_1), \quad (x_2, y_2, z_2), \quad (x_3, y_3, z_3)$$

$$\begin{vmatrix} x - x_1 & y - y_1 & z - z_1 \\ x_2 - x_1 & y_2 - y_1 & z_2 - z_1 \\ x_3 - x_1 & y_3 - y_1 & z_3 - z_1 \end{vmatrix} = 0$$

or

$$\begin{vmatrix} y_2 - y_1 & z_2 - z_1 \\ y_3 - y_1 & z_3 - z_1 \end{vmatrix}(x - x_1) + \begin{vmatrix} z_2 - z_1 & x_2 - x_1 \\ z_3 - z_1 & x_3 - x_1 \end{vmatrix}(y - y_1) + \begin{vmatrix} x_2 - x_1 & y_2 - y_1 \\ x_3 - x_1 & y_3 - y_1 \end{vmatrix}(z - z_1) = 0$$

## EQUATION OF PLANE IN INTERCEPT FORM

$$\frac{x}{a} + \frac{y}{b} + \frac{z}{c} = 1$$

where $a$, $b$, $c$ are the intercepts on the $x$, $y$, $z$ axes, respectively.

## EQUATIONS OF LINE THROUGH $(x_0, y_0, z_0)$ AND PERPENDICULAR TO PLANE

$$Ax + By + Cz + D = 0$$

$$\frac{x - x_0}{A} = \frac{y - y_0}{B} = \frac{z - z_0}{C}$$

$$\text{or} \quad x = x_0 + At, \quad y = y_0 + Bt, \quad z = z_0 + Ct$$

## DISTANCE FROM POINT $(x, y, z)$ TO PLANE $Ax + By + D = 0$

$$\frac{Ax_0 + By_0 + Cz_0 + D}{\pm\sqrt{A^2 + B^2 + C^2}}$$

where the sign is chosen so that the distance is nonnegative.

## NORMAL FORM FOR EQUATION OF PLANE

$$x\cos\alpha + y\cos\beta + z\cos\gamma = p$$

where $p$ is the perpendicular distance from $O$ to plane at $P$ and $\alpha, \beta, \gamma$ are angles between $OP$ and positive $x, y, z$ axes.

## TRANSFORMATION OF COORDINATES INVOLVING PURE TRANSLATION

$$\begin{cases} x = x' + x_0 \\ y = y' + y_0 \\ z = z' + z_0 \end{cases} \quad \text{or} \quad \begin{cases} x' = x + x_0 \\ y' = y + y_0 \\ z' = z + z_0 \end{cases}$$

where $(x, y, z)$ are old coordinates relative to the $xyz$ system, $(x', y', z')$ are new coordinates relative to the $(x', y', z')$ system, and $(x_0, y_0, z_0)$ are the coordinates of the new origin $O'$ relative to the old $xyz$ coordinate system.

## TRANSFORMATION OF COORDINATES INVOLVING PURE ROTATION

$$\begin{cases} x = l_1 x' + l_2 y' + l_3 z' \\ y = m_1 x' + m_2 y' + m_3 z' \\ z = n_1 x' + n_2 y' + n_3 z' \end{cases} \quad \text{or} \quad \begin{cases} x' = l_1 x + m_1 y + n_1 z \\ y' = l_2 x + m_2 y + n_3 z \\ z' = l_3 x + m_3 y + n_3 z \end{cases}$$

where the origins of the $xyz$ and $x', y', z'$ systems are the same and $l_1, m_1, n_1$; $l_2, m_2, n_2$; $l_3, m_3, n_3$ are the direction cosines of the $x', y', z'$ axes relative to the $x, y, z$ axes, respectively.

## TRANSFORMATION OF COORDINATES INVOLVING TRANSLATION AND ROTATION

$$\begin{cases} x = l_1 x' + l_2 y' + l_3 z' + x_0 \\ y = m_1 x' + m_2 y' + m_3 z' + y_0 \\ z = n_1 x' + n_2 y' + n_3 z' + z_0 \end{cases}$$

or

$$\begin{cases} x' = l_1(x - x_0) + m_1(y - y_0) + n_1(z - z_0) \\ y' = l_2(x - x_0) + m_2(y - y_0) + n_2(z - z_0) \\ z' = l_3(x - x_0) + m_3(y - y_0) + n_3(z - z_0) \end{cases}$$

where the origin $O'$ of the $x'$ $y'$ $z'$ system has coordinates $(x_0, y_0, z_0)$ relative to the $xyz$ system and $l_1, m_1, n_1$; $l_2, m_2, n_2$; $l_3, m_3, n_3$ are the direction cosines of the $x'$ $y'$ $z'$ axes relative to the $x, y, z$ axes, respectively.

## Cylindrical Coordinates $(r, \theta, z)$

A point P can be located by cylindrical coordinates $(r, \theta, z)$ as well as rectangular coordinates $(x, y, z)$. The transformation between these coordinates is

$$\begin{cases} x = r\cos\theta \\ y = r\sin\theta \\ z = z \end{cases} \quad \text{or} \quad \begin{cases} r = \sqrt{x^2 + y^2} \\ \theta = \tan^{-1}(y/x) \\ z = z \end{cases}$$

## Spherical Coordinates $(r, \theta, \phi)$

A point P can be located by cylindrical coordinates $(r, \theta, \phi)$ as well as rectangular coordinates $(x, y, z)$. The transformation between these coordinates is

$$\begin{cases} x = r\cos\theta\cos\phi \\ y = r\sin\theta\sin\phi \\ z = r\cos\theta \end{cases} \quad \text{or} \quad \begin{cases} r = \sqrt{x^2 + y^2 + z^2} \\ \phi = \tan^{-1}(y/x) \\ \theta = \cos^{-1}\left(z/\sqrt{x^2 + y^2 + z^2}\right) \end{cases}$$

Equation of sphere in rectangular coordinates

$$(x - x_0)^2 + (y - y_0)^2 + (z - z_0)^2 = R^2$$

where the sphere has cent $(x_0, y_0, z_0)$ and radius R.

Equation of sphere in cylindrical coordinates

$$r^2 - 2r_0 r\cos(\theta - \theta_0) + r_0^2 + (z - z_0)^2 = R^2$$

where the sphere has center $(r_0, \theta_0, z_0)$ in cylindrical coordinates and radius R.

If the center is at the origin, the equation is

$$r^2 + z^2 = R^2$$

Equation of sphere in spherical coordinates

$$r^2 + r_0^2 - 2r_0 r\sin\theta\sin\theta_0 \cos(\phi - \phi_0) = R^2$$

where the sphere has center $(r_0, \theta_0, \phi_0)$ in spherical coordinates and radius R.

If the center is at the origin, the equation is

$$r = R$$

## Logarithmic Identities

$$\text{Ln}(z_1 z_2) = \text{Ln } z_1 + \text{Ln } z_2$$

$$\ln(z_1 z_2) = \ln z_1 + \ln z_2 \quad (-\pi < \arg z_1 + \arg z_2 \leq \pi)$$

$$\text{Ln } \frac{z_1}{z_2} = \text{Ln } z_1 - \text{Ln } z_2$$

$$\ln \frac{z_1}{z_2} = \ln z_1 - \ln z_2 \quad (-\pi < \arg z_1 - \arg z_2 \leq \pi)$$

$$\text{Ln } z^n = n \text{ Ln } z \quad (n \text{ integer})$$

$$\ln z^n = n \ln z \quad (n \text{ integer}, \; -\pi < n \arg z \leq \pi)$$

## Special Values

$$\ln 1 = 0$$

$$\ln 0 = -\infty$$

$$\ln(-1) = \pi i$$

$$\ln(\pm i) = \pm \frac{1}{2}\pi i$$

$\ln e = 1$, $e$ is the real number such that

$$\int_1^e \frac{dt}{t} = 1$$

$$e = \lim_{n \to \infty} \left(1 + \frac{1}{n}\right)^n = 2.71828 \quad 18284\ldots$$

## Logarithms to General Base

$$\log_a z = \ln z / \ln a$$

$$\log_a z = \frac{\log_b z}{\log_b a}$$

$$\log_a b = \frac{1}{\log_b a}$$

$$\log_e z = \ln z$$

$$\log_{10} z = \ln z / \ln 10 = \log_{10} e \ln z = (0.43429 \quad 44819\ldots) \ln z$$

$$\ln z = \ln 10 \log_{10} z = (2.30258 \quad 50929\ldots) \log_{10} z$$

# Basic Mathematical Calculations

$$\begin{pmatrix} \log_e x = \ln x, \text{ called natural, Napierian, or hyperbolic logarithms} \\ \log_{10} x, \text{ called common or Briggs logarithms} \end{pmatrix}$$

## Series Expansions

$$\ln(1+z) = z - \frac{1}{2}z^2 + \frac{1}{3}z^3 - \ldots \left(|z| \leq 1 \text{ and } z \neq -1\right)$$

$$\ln z = \left(\frac{z-1}{z}\right) + \frac{1}{2}\left(\frac{z-1}{z}\right)^2 + \frac{1}{3}\left(\frac{z-1}{z}\right)^3 + \ldots \left(\Re z \geq \frac{1}{2}\right)$$

$$\ln z = (z-1) - \frac{1}{2}(z-1)^2 + \frac{1}{3}(z-1)^3 - \ldots \left(|z-1| \leq 1, \quad z \neq 0\right)$$

$$\ln z = 2\left[\left(\frac{z-1}{z+1}\right) + \frac{1}{3}\left(\frac{z-1}{z+1}\right)^3 + \frac{1}{5}\left(\frac{z-1}{z+1}\right)^5 + \ldots\right] \left(\Re z \geq 0, \quad z \neq 0\right)$$

$$\ln\left(\frac{z+1}{z-1}\right) = 2\left(\frac{1}{z} + \frac{1}{3z^3} + \frac{1}{5z^5} + \ldots\right) \left(|z| \geq 1, \quad z \neq \pm 1\right)$$

$$\ln(z+a) = \ln a + 2\left[\left(\frac{z}{2a+z}\right) + \frac{1}{3}\left(\frac{z}{2a+z}\right)^3 + \frac{1}{5}\left(\frac{z}{2a+z}\right)^5 + \ldots\right] \left(a > 0, \quad \Re z \geq -a \neq z\right)$$

## Limiting Values

$$\lim_{x \to \infty} x^{-\alpha} \ln x = 0 \quad \left(\alpha \text{ constant}, \quad \Re\alpha > 0\right)$$

$$\lim_{x \to 0} x^{\alpha} \ln x = 0 \quad \left(\alpha \text{ constant}, \quad \Re\alpha > 0\right)$$

$$\lim_{m \to \infty}\left(\sum_{k=1}^{m} \frac{1}{k} - \ln m\right) = \gamma \quad (\text{Euler's constant}) = .57721\ 56649\ldots$$

## Inequalities

$$\frac{x}{1+x} < \ln(1+x) < x \quad (x > -1, \quad x \neq 0)$$

$$x < -\ln(1-x) < \frac{x}{1+x} \quad (x < 1, \quad x \neq 0)$$

$$|\ln(1-x)| < \frac{3x}{2} \quad (0 < x \leq .5828)$$

$$\ln x \le x - 1 \quad (x > 0)$$

$$\ln x \le n\left(x^{1/n} - 1\right) \text{ for any positive } n \quad (x > 0)$$

$$|\ln(1-z)| \le -\ln(1-|z|) \quad (|z| < 1)$$

## CONTINUED FRACTIONS

$$\ln(1+z) = \frac{z}{1+}\frac{z}{2+}\frac{z}{3+}\frac{4z}{4+}\frac{4z}{5+}\frac{9z}{6+}\cdots$$

$$(z \text{ in the plane cut from} -1 \text{ to} -\infty)$$

$$\ln\left(\frac{1+z}{1-z}\right) = \frac{2z}{1-}\frac{z^2}{3-}\frac{4z^2}{5-}\frac{9z^2}{7-}\cdots$$

## POLYNOMIAL APPROXIMATIONS

$$\frac{1}{\sqrt{10}} \le x \le \sqrt{10}$$

$$\log_{10} x = a_1 t + a_3 t^3 + \varepsilon(x), \quad t = (x-1)/(x+1)$$

$$|\varepsilon(x)| \le 6 \times 10^{-4}$$

$$a_1 = .86304 \quad a_3 = .36415$$

$$\frac{1}{\sqrt{10}} \le x \le \sqrt{10}$$

$$\log_{10} x = a_1 t + a_3 t^3 + a_5 t^5 + a_7 t^7 + a_9 t^9 + \varepsilon(x)$$

$$t = (x-1)/(x+1)$$

$$|\varepsilon(x)| \le 10^{-7}$$

$$a_1 = .86859\ 1718$$

$$a_3 = .28933\ 5524$$

$$a_5 = .17752\ 2071$$

$$a_7 = .09437\ 6476$$

$$a_9 = .19133\ 7714$$

# Basic Mathematical Calculations

$$0 \leq x \leq 1$$

$$\ln(1+x) = a_1 x + a_2 x^2 + a_3 x^3 + a_4 x^4 + a_5 x^5 + \varepsilon(x)$$

$$|\varepsilon(x)| \leq 1 \times 10^{-5}$$

$$a_1 = .99949\ 556$$

$$a_2 = .49190\ 896$$

$$a_3 = .28947\ 478$$

$$a_4 = .13606\ 275$$

$$a_5 = .03215\ 845$$

$$0 \leq x \leq 1$$

$$\ln(1+x) = a_1 x + a_2 x^2 + a_3 x^3 + a_4 x^4 + a_5 x^5 + a_6 x^6 + a_7 x^7 + a_8 x^8 + \varepsilon(x)$$

$$|\varepsilon(x)| \leq 3 \times 10^{-8}$$

$$a_1 = .99999\ 64239$$

$$a_2 = -.49987\ 41238$$

$$a_3 = .33179\ 90258$$

$$a_4 = -.24073\ 38084$$

$$a_5 = .16765\ 40711$$

$$a_6 = -.09532\ 93897$$

$$a_7 = .03608\ 84937$$

$$a_8 = -.00645\ 35442$$

Exponential function series expansion

$$e^z = \exp z = 1 + \frac{z}{1!} + \frac{z^2}{2!} + \frac{z^3}{3!} + \ldots \quad (z = x + iy)$$

## FUNDAMENTAL PROPERTIES

$$\mathrm{Ln}(\exp z) = z + 2k\pi i \quad (k \text{ any integer})$$

$$\ln(\exp z) = z \qquad \left(-\pi < \oint z \le \pi\right)$$

$$\exp(\ln z) = \exp(\operatorname{Ln} z) = z$$

$$\frac{d}{dz}\exp z = \exp z$$

## DEFINITION OF GENERAL POWERS

$$\text{If } N = a^z, \text{ then } z = \log_a N$$

$$a^z = \exp(z \ln a)$$

$$\text{If } a = |a|\exp(i \arg a) \qquad (-\pi < \arg a \le \pi)$$

$$|a^z| = |a|^x e^{-y \arg a}$$

$$\arg(a^z) = y \ln|a| + x \arg a$$

$$\operatorname{Ln} a^z = z \ln a \qquad \text{for one of the values of } \operatorname{Ln} a^z$$

$$\ln a^x = x \ln a \qquad (a \text{ real and positive})$$

$$|e^z| = e^x$$

$$\arg(e^z) = y$$

$$a^{z_1} a^{z_2} = a^{z_1 + z_2}$$

$$a^z b^z = (ab)^z \qquad (-\pi < \arg a + \arg b \le \pi)$$

## LOGARITHMIC AND EXPONENTIAL FUNCTIONS

Periodic property

$$e^{z + 2\pi k i} = e^z \qquad (k \text{ any integer})$$

$$e^x < \frac{1}{1-x} \qquad (x < 1)$$

$$\frac{1}{1-x} < (1 - e^{-x}) < x \qquad (x > -1)$$

$$x < (e^x - 1) < \frac{1}{1-x} \qquad (x < 1)$$

# Basic Mathematical Calculations

$$1 + x > e^{\frac{x}{1+x}} \quad (x > -1)$$

$$e^x > 1 + \frac{x^n}{n!} \quad (n > 0, \ x > 0)$$

$$e^x > \left(1 + \frac{x}{y}\right)^y > \frac{xy}{e^{x+y}} \quad (x > 0, \ y > 0)$$

$$e^{-x} < 1 - \frac{x}{2} \quad (0 < x \leq 1.5936)$$

$$\frac{1}{4}|z| < |e^z - 1| < \frac{7}{4}|z| \quad (0 < |z| < 1)$$

$$|e^z - 1| \leq e^{|z|} - 1 \leq |z|e^{|z|} \quad (\text{all } z)$$

$$e^{2a \arctan \frac{1}{z}} = 1 + \frac{2a}{z-a+} \frac{a^2+1}{3z+} \frac{a^2+4}{5z+} \frac{a^2+9}{7z+} \cdots$$

## Polynomial Approximations

$$0 \leq x \leq \ln 2 = .693\ldots$$

$$e^{-x} = 1 + a_1 x + a_2 x^2 + \varepsilon(x)$$

$$|\varepsilon(x)| \leq 3 \times 10^{-3}$$

$$a_1 = -.9664$$

$$a_2 = .3536$$

$$0 \leq x \leq \ln 2$$

$$e^{-x} = 1 + a_1 x + a_2 x^2 + a_3 x^3 + a_4 x^4 + \varepsilon(x)$$

$$|\varepsilon(x)| \leq 3 \times 10^{-5}$$

$$a_1 = -.99986 \ \ 84$$

$$a_2 = .49829 \ \ 26$$

$$a_3 = -.15953 \ \ 32$$

$$a_4 = .02936 \quad 41$$

$$0 \leq x \leq \ln 2$$

$$e^{-x} = 1 + a_1 x + a_2 x^2 + a_3 x^3 + a_4 x^4 + a_5 x^5 + a_6 x^6 + a_7 x^7 + \varepsilon(x)$$

$$|\varepsilon(x)| \leq 2 \times 10^{-10}$$

$$a_1 = -.99999 \quad 99995$$

$$a_2 = .49999 \quad 99206$$

$$a_3 = -.16666 \quad 53019$$

$$a_4 = .04165 \quad 73475$$

$$a_5 = -.00830 \quad 13598$$

$$a_6 = .00132 \quad 98820$$

$$a_7 = -.00014 \quad 13161$$

$$0 \leq x \leq 1$$

$$10^x = \left(1 + a_1 x + a_2 x^2 + a_3 x^3 + a_4 x^4\right)^2 + \varepsilon(x)$$

$$|\varepsilon(x)| \leq 7 \times 10^{-4}$$

$$a_1 = 1.14991 \quad 96$$

$$a_2 = .67743 \quad 23$$

$$a_3 = .20800 \quad 30$$

$$a_4 = .12680 \quad 89$$

$$0 \leq x \leq 1$$

$$10^x = \left(1 + a_1 x + a_2 x^2 + a_3 x^3 + a_4 x^4 + a_5 x^5 + a_6 x^6 + a_7 x^7\right)^2 + \varepsilon(x)$$

$$|\varepsilon(x)| \leq 5 \times 10^{-8}$$

$$a_1 = 1.15129 \quad 277603$$

$$a_2 = .66273 \quad 088429$$

$$a_3 = .25439 \quad 357484$$

$$a_4 = .07295 \quad 173666$$

# Basic Mathematical Calculations

$$a_5 = .01742 \ 111988$$
$$a_6 = .00255 \ 491796$$
$$a_7 = .00093 \ 264267$$

Tables 4.1–4.7 show functional relationships for the equations and formulas that follow.

## TABLE 4.1
### Basic Laplace Transforms

**Laplace Transforms of Some Basic Functions**

| $f(t)$ | $\mathcal{L}\{f(t)\} = F(s)$ |
| --- | --- |
| 1 | $\dfrac{1}{s}$ |
| $t^n, \quad n = 1, 2, 3, \ldots$ | $\dfrac{n!}{s^{n+1}}$ |
| $e^{at}$ | $\dfrac{1}{s-a}$ |
| $\sin kt$ | $\dfrac{k}{s^2 + k^2}$ |
| $\cos kt$ | $\dfrac{s}{s^2 + k^2}$ |
| $\sinh kt$ | $\dfrac{k}{s^2 - k^2}$ |
| $\cosh kt$ | $\dfrac{s}{s^2 - k^2}$ |

## TABLE 4.2
### Operational Properties of Transforms

| | |
| --- | --- |
| $e^{at} f(t)$ | $F(s-a)$ |
| $f(t-a)\,\mathcal{u}(t-a), a > 0$ | $e^{-as} F(s)$ |
| $t^n f(t), \quad n = 1, 2, 3, \ldots$ | $(-1)^n \dfrac{d^n}{ds^n} F(s)$ |
| $f^{(n)}(t), \quad n = 1, 2, 3, \ldots$ | $s^n F(s) - s^{n-1} f(0) - \cdots - f^{(n-1)}(0)$ |
| $\displaystyle\int_0^t f(\tau)\,d\tau$ | $\dfrac{F(s)}{s}$ |
| $\displaystyle\int_0^t f(\tau) g(t - \tau)\,d\tau$ | $F(s) G(s)$ |

## TABLE 4.3
## Transforms of Functional Products

| | |
|---|---|
| $t^n e^{at}, \quad n = 1, 2, 3, \ldots$ | $\dfrac{n!}{(s-a)^{n+1}}$ |
| $e^{at} \sin kt$ | $\dfrac{k}{(s-a)^2 + k^2}$ |
| $e^{at} \cos kt$ | $\dfrac{s-a}{(s-a)^2 + k^2}$ |
| $t \sin kt$ | $\dfrac{2ks}{\left(s^2 + k^2\right)^2}$ |
| $t \cos kt$ | $\dfrac{s^2 - k^2}{\left(s^2 + k^2\right)^2}$ |
| $\sin kt - kt \cos kt$ | $\dfrac{2k^3}{\left(s^2 + k^2\right)^2}$ |
| $\sin kt + kt \cos kt$ | $\dfrac{2ks^2}{\left(s^2 + k^2\right)^2}$ |

## TABLE 4.4
## Units of Measurement

**English System**

| | | |
|---|---|---|
| 1 foot (ft) | = 12 inches (in) | $1' = 12''$ |
| 1 yard (yd) | = 3 feet | |
| 1 mile (mi) | = 1760 yards | |
| 1 sq. foot | = 144 sq. inches | |
| 1 sq. yard | = 9 sq. feet | |
| 1 acre | = 4840 sq. yard = 43560 ft$^2$ | |
| 1 sq. mile | = 640 acres | |

**Metric System**

| | | |
|---|---|---|
| mm | millimeter | .001 m |
| cm | centimeter | .01 m |
| dm | decimeter | .1 m |
| m | meter | 1 m |
| dam | decameter | 10 m |
| hm | hectometer | 100 m |
| km | kilometer | 1000 m |

# Basic Mathematical Calculations

## TABLE 4.5
## Common Units of Measurement

### Common Units used with the International System

| Units of Measurement | Abbreviation | Relation | Units of Measurement | Abbreviation | Relation |
|---|---|---|---|---|---|
| meter | m | Length | degree Celsius | °C | Temperature |
| hectare | ha | Area | kelvin | K | Thermodynamic temperature |
| tonne | t | Mass | pascal | Pa | Pressure, stress |
| kilogram | kg | Mass | joule | J | Energy, work |
| nautical mile | M | Distance (navigation) | newton | N | Force |
| knot | kn | Speed (navigation) | watt | W | Power, radiant flux |
| liter | L | Volume or capacity | ampere | A | Electric current |
| second | s | Time | volt | V | Electric potential |
| hertz | Hz | Frequency | ohm | Ω | Electric resistance |
| candela | cd | Luminous intensity | coulomb | C | Electric charge |

## TABLE 4.6
## Values of Trig Ratio

| $\theta$ | 0 | $\pi/2$ | $\pi$ | $3\pi/2$ | $2\pi$ |
|---|---|---|---|---|---|
| $\sin\theta$ | 0 | 1 | 0 | −1 | 0 |
| $\cos\theta$ | 1 | 0 | −1 | 0 | 1 |
| $\tan\theta$ | 0 | $\infty$ | 0 | $-\infty$ | 0 |

## TABLE 4.7
## Physics Equations

$D = \dfrac{m}{V}$     $D$ density  
        $m$ mass $\left(\dfrac{g}{cm^3} = \dfrac{kg}{m^3}\right)$  
        $V$ volume

$d = vt$    $d$ distance m  
       $v$ velocity m/s  
       $t$ time s

$p = \dfrac{W}{t}$    $P$ power W (=watts)  
       $W$ work J  
       $t$ time s

$KE = \dfrac{1}{2}mv^2$    $KE$ kinetic energy J  
            $m$ mass kg  
            $v$ velocity m/s

*(Continued)*

## TABLE 4.7 (*Continued*)
## Physics Equations

| | | | |
|---|---|---|---|
| $a = \dfrac{vf - vi}{t}$ | $a$ acceleration m/s² <br> $vf$ final velocity m/s <br> $vi$ initial velocity m/s <br> $t$ time s | $Fe = \dfrac{kQ_1Q_2}{d^2}$ | $Fe$ electrical force N <br> $k$ Coulomb's constant <br> $\left(k = 9 \times 10^9 \dfrac{Nm^2}{c^2}\right)$ <br> $Q_1, Q_2$ are electrical charges <br> $d$ separation distance m |
| $d = vit + \dfrac{1}{2}at^2$ | $d$ distance m <br> $vi$ initial velocity m/s <br> $t$ time s <br> $a$ acceleration m/s² | $V = \dfrac{W}{Q}$ | $V$ electrical potential difference V (=volts) <br> $W$ work done J <br> $Q$ electric charge |
| $F = ma$ | $F$ net force N (=newtons) <br> $m$ mass kg <br> $a$ acceleration m/s² | $I = \dfrac{Q}{t}$ | $I$ electric current amperes <br> $Q$ electric charge flowing C <br> $t$ time s |
| $Fg = \dfrac{Gm_1m_2}{d^2}$ | $Fg$ force of gravity N <br> $G$ universal gravitational constant $\left(G = 6.67 \times 10^{-11} \dfrac{N - m^2}{kg^2}\right)$ <br> $m_1, m_2$ masses of the two objects kg <br> $d$ separation distance m | $W = VIt$ | $W$ electrical energy J <br> $V$ current V <br> $I$ current A <br> $t$ time s |
| $p = mv$ | $p$ momentum kgm/s <br> $m$ mass kg <br> $v$ velocity m/s | $p = VI$ | $P$ power W <br> $V$ voltage V <br> $I$ current A |
| $W = Fd$ | $W$ work J (=joules) <br> $F$ force N <br> $d$ distance m | $H = cm\Delta T$ | $H$ heat energy J <br> $m$ mass kg <br> $\Delta T$ change in temperature °C <br> $c$ specific heat J/Kg°C |

Surface area of cylinder = $2\pi rh + 2\pi r^2$
Volume of cylinder = $\pi r^2 h$
Figures 4.7–4.11 illustrate the shapes related to the formula measures that follow.
Surface area of a cone = $\pi r^2 + \pi rs$
Volume of a cone = $\dfrac{\pi r^2 h}{3}$
Volume of a pyramid = $\dfrac{Bh}{3}$
($B$ = area of base)

Basic Mathematical Calculations

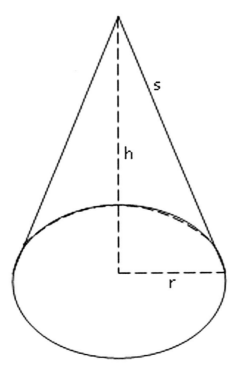

**FIGURE 4.7** Surface area of a cone.

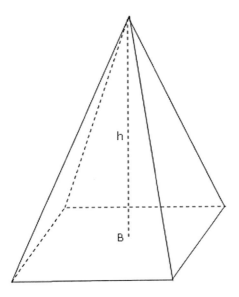

**FIGURE 4.8** Volume of a pyramid.

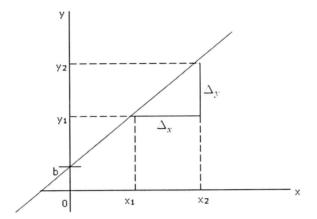

**FIGURE 4.9** Equation of a straight line.

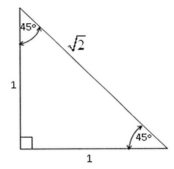

**FIGURE 4.10** Triangle equations.

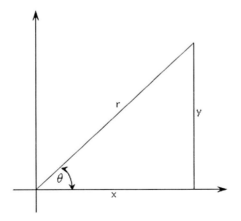

**FIGURE 4.11** Right triangle calculations.

## SLOPES

Equation of a straight line: $y - y_1 = m(x - x_1)$

where $m = \text{slope} = \dfrac{\text{rise}}{\text{run}}$

$$= \frac{\Delta y}{\Delta x} = \frac{y_2 - y_1}{x_2 - x_1}$$

or

$$y = mx + b$$

where

$$m = \text{slope}$$
$$b = y\text{-intercept}$$

## TRIGONOMETRIC RATIOS

$$\tan\theta = \frac{\sin\theta}{\cos\theta}$$

$$\sin^2\theta + \cos^2\theta = 1$$

$$1 + \tan^2\theta = \sec^2\theta$$

$$1 + \cot^2\theta = \csc^2\theta$$

$$\cos^2\theta - \sin^2\theta = \cos 2\theta$$

$$\sin 45° = \frac{1}{\sqrt{2}}$$

$$\cos 45° = \frac{1}{\sqrt{2}}$$

$$\tan 45° = 1$$

$$\sin(A + B) = \sin A \cos B + \cos A \sin B$$

$$\sin(A - B) = \sin A \cos B - \cos A \sin B$$

$$\cos(A + B) = \cos A \cos B - \sin A \sin B$$

$$\cos(A - B) = \cos A \cos B + \sin A \sin B$$

$$\tan(A+B) = \frac{\tan A + \tan B}{1 - \tan A \tan B}$$

$$\tan(A-B) = \frac{\tan A - \tan B}{1 + \tan A \tan B}$$

$$\sin\theta = \frac{y}{r}(\text{opposite/hypotenuse}) = 1/\csc\theta$$

$$\cos\theta = \frac{x}{r}(\text{adjacent/hypotenuse}) = 1/\sec\theta$$

$$\tan\theta = \frac{y}{x}(\text{opposite/adjacent}) = 1/\cot\theta$$

Figures 4.12 and 4.13 illustrate angle measurements.

$$\sin 30° = \frac{1}{2} \qquad \sin 60° = \frac{\sqrt{3}}{2}$$

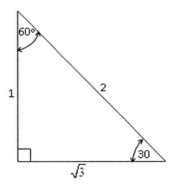

**FIGURE 4.12** Angle relationships.

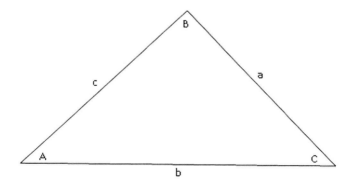

**FIGURE 4.13** Cosine law.

# Basic Mathematical Calculations

$$\cos 30° = \frac{\sqrt{3}}{2} \qquad \cos 60° = \frac{1}{2}$$

$$\tan 30° = \frac{1}{\sqrt{3}} \qquad \tan 60° = \sqrt{3}$$

## Sine Law

$$\frac{a}{\sin A} = \frac{b}{\sin B} = \frac{c}{\sin C}$$

## Cosine Law

$$a^2 = b^2 + c^2 - 2bc \cos A$$

$$b^2 = a^2 + c^2 - 2ac \cos B$$

$$c^2 = a^2 + b^2 - 2ab \cos C$$

$$\theta = 1 \text{ radian}$$

$$2\pi \text{ radians} = 360°$$

# ALGEBRA

## Expanding

$$a(b+c) = ab + ac$$

$$(a+b)^2 = a^2 + 2ab + b^2$$

$$(a-b)^2 = a^2 - 2ab + b^2$$

$$(a+b)(c+d) = ac + ad + bc + bd$$

$$(a+b)^3 = a^3 + 3a^2b + 3ab^2 + b^3$$

$$(a-b)^3 = a^3 - 3a^2b + 3ab^2 - b^3$$

## Factoring

$$a^2 - b^2 = (a+b)(a-b)$$

$$a^2 + 2ab + b^2 = (a+b)^2$$

$$a^3 + b^3 = (a+b)(a^2 - ab + b^2)$$

$$a^3b - ab = ab(a+1)(a-1)$$

$$a^2 - 2ab + b^2 = (a-b)^2$$
$$a^3 - b^3 = (a-b)(a^2 + ab + b^2)$$

## ROOTS OF QUADRATIC

The solution for isa quadratic equation $ax^2 + bx + c = 0$ is

$$x = \frac{-b \pm \sqrt{b^2 - 4ac}}{2a}$$

## LAW OF EXPONENTS

$$a^r \cdot a^s = a^{r+s}$$
$$\frac{a^p a^q}{a^r} = a^{p+q-r}$$
$$\frac{a^r}{a^s} = a^{r-s}$$
$$(a^r)^s = a^{rs}$$
$$(ab)^r = a^r b^r$$
$$\left(\frac{a}{b}\right)^r = \frac{a^r}{b^r} (b \neq 0)$$
$$a^0 = 1 (a \neq 0)$$
$$a^{-r} = \frac{1}{a^r} (a \neq 0)$$
$$a^{\frac{r}{s}} = \sqrt[s]{a^r} \qquad a^{\frac{1}{2}} = \sqrt{a} \qquad a^{\frac{1}{3}} = \sqrt[3]{a}$$

## LOGARITHMS

Example:

$$\log(xy) = \log x + \log y \qquad \log\left(\frac{x}{y}\right) = \log x - \log y$$
$$\log x^r = r \log x$$
$$\log x = n \leftrightarrow x = 10^n \text{ (common log)} \qquad \pi \approx 3.14159265$$
$$\log_a x = n \leftrightarrow x = a^n \text{ (log to the base } a) \qquad e \approx 2.71828183$$
$$\ln x = n \leftrightarrow x = e^n \text{ (natural log)}$$

# 5 Statistical Methods for Data Analytics

Statistics is the language of data.

## INTRODUCTION

Data analytics relies on statistical distributions and methods. The most common distributions are summarized in Table 5.1 (Leemis, 1987).

## DISCRETE DISTRIBUTIONS

Probability mass function, $p(x)$
  Mean, $\mu$

  Variance, $\sigma^2$
  Coefficient of skewness, $\beta_1$
  Coefficient of kurtosis, $\beta_2$
  Moment-generating function, $M(t)$
  Characteristic function, $\phi(t)$
  Probability-generating function, $P(t)$

### BERNOULLI DISTRIBUTION

$$p(x) = p^x q^{x-1} \quad x = 0,1 \quad 0 \leq p \leq 1 \quad q = 1-p$$

$$\mu = p \quad \sigma^2 = pq \quad \beta_1 = \frac{1-2p}{\sqrt{pq}} \quad \beta_2 = 3 + \frac{1-6pq}{pq}$$

$$M(t) = q + pe^t \quad \phi(t) = q + pe^{it} \quad P(t) = q + pt$$

### BETA BINOMIAL DISTRIBUTION

$$p(x) = \frac{1}{n+1} \frac{B(a+x, b+n-x)}{B(x+1, n-x+1) B(a,b)} \quad x = 0,1,2,\ldots,n \quad a > 0 \quad b > 0$$

## TABLE 5.1
### Summary of Common Statistical Distributions

| Distribution of Random Variable x | Functional Form | Parameters | Mean | Variance | Range |
|---|---|---|---|---|---|
| Binomial | $P_x(k) = \dfrac{n!}{k!(n-k)!} p^k (1-p)^{n-k}$ | $n, p$ | $np$ | $np(1-p)$ | $0, 1, 2, \ldots, n$ |
| Poisson | $P_x(k) = \dfrac{\lambda^k e^{-\lambda}}{k!}$ | $\lambda$ | $\lambda$ | $\lambda$ | $0, 1, 2, \ldots$ |
| Geometric | $P_x(k) = p(1-p)^{k-1}$ | $p$ | $1/p$ | $\dfrac{1-p}{p^2}$ | $1, 2, \ldots$ |
| Exponential | $f_x(y) = \dfrac{1}{\theta} e^{-y/\theta}$ | $\theta$ | $\theta$ | $\theta^2$ | $(0, \infty)$ |
| Gamma | $f_x(y) = \dfrac{1}{\Gamma(\alpha)\beta^\alpha} y^{(\alpha-1)} e^{-y/\beta}$ | $\alpha, \beta$ | $\alpha\beta$ | $\alpha\beta^2$ | $(0, \infty)$ |

*(Continued)*

## TABLE 5.1 (Continued)
### Summary of Common Statistical Distributions

| Distribution of Random Variable x | Functional Form | Parameters | Mean | Variance | Range |
|---|---|---|---|---|---|
| Beta | $f_x(y) = \dfrac{\Gamma(\alpha+\beta)}{\Gamma(\alpha)\Gamma(\beta)} y^{(\alpha-1)}(1-y)^{(\beta-1)}$ | $\alpha, \beta$ | $\dfrac{\alpha}{\alpha+\beta}$ | $\dfrac{\alpha\beta}{(\alpha+\beta)^2(\alpha+\beta+1)}$ | $(0, 1)$ |
| Normal | $f_x(y) = \dfrac{1}{\sqrt{2\pi}\sigma} e^{-(y-\mu)^2/2\sigma^2}$ | $\mu, \sigma$ | $\mu$ | $\sigma^2$ | $(-\infty, \infty)$ |
| Student t | $f_x(y) = \dfrac{1}{\sqrt{\pi v}} \dfrac{\Gamma\left(\dfrac{v+1}{2}\right)}{\Gamma(v/2)} (1+y^2/v)^{-(v+1)/2}$ | $v$ | $0$ for $v>1$ | $\dfrac{v}{v-2}$ for $v>2$ | $(-\infty, \infty)$ |
| Chi-square | $f_x(y) = \dfrac{1}{2^{v/2}\Gamma(v/2)} y^{(v-2)/2} e^{-y/2}$ | $v$ | $v$ | $2v$ | $(0, \infty)$ |
| F | $f_x(y) = \dfrac{\Gamma\left(\dfrac{v_1+v_2}{2}\right)}{\Gamma\left(\dfrac{v_1}{2}\right)\Gamma\left(\dfrac{v_2}{2}\right)} v_1^{v_1/2} v_2^{v_2/2} \dfrac{(y)^{(v_1/2)-1}}{(v_2+v_1 y)^{\frac{v_1+v_2}{2}}}$ | $v_1, v_2$ | $\dfrac{v_2}{v_2-2}$ for $v_2>2$ | $\dfrac{v_2^2(2v_2+2v_1-4)}{v_1(v_2-2)^2(v_2-4)}$ for $v_2>4$ | $(0, \infty)$ |

$$\mu = \frac{na}{a+b} \qquad \sigma^2 = \frac{nab(a+b+n)}{(a+b)^2(a+b+1)} \qquad B(a,b) \quad \text{is the beta function.}$$

## Beta Pascal Distribution

$$p(x) = \frac{\Gamma(x)\Gamma(v)\Gamma(\rho+v)\Gamma(v+x-(\rho+r))}{\Gamma(r)\Gamma(x-r+1)\Gamma(\rho)\Gamma(v-\rho)\Gamma(v+x)} \qquad x = r, r+1, \ldots \quad v > \rho > 0$$

$$\mu = r\frac{v-1}{\rho-1}, \; \rho > 1 \qquad \sigma^2 = r(r+\rho-1)\frac{(v-1)(v-\rho)}{(\rho-1)^2(\rho-2)}, \; \rho > 2$$

## Binomial Distribution

$$p(x) = \binom{n}{x} p^x q^{n-x} \quad x = 0, 1, 2, \ldots, n \quad 0 \leq p \leq 1 \quad q = 1-p$$

$$\mu = np \qquad \sigma^2 = npq \qquad \beta_1 = \frac{1-2p}{\sqrt{npq}} \qquad \beta_2 = 3 + \frac{1-6pq}{npq}$$

$$M(t) = (q+pe^t)^n \qquad \phi(t) = (q+pe^{it})^n \qquad P(t) = (q+pt)^n$$

## Discrete Weibull Distribution

$$p(x) = (1-p)^{x^\beta} - (1-p)^{(x+1)^\beta} \qquad x = 0, 1, \ldots \quad 0 \leq p \leq 1 \quad \beta > 0$$

## Geometric Distribution

$$p(x) = pq^{1-x} \quad x = 0, 1, 2, \ldots \quad 0 \leq p \leq 1 \quad q = 1-p$$

$$\mu = \frac{1}{p} \qquad \sigma^2 = \frac{q}{p^2} \qquad \beta_1 = \frac{2-p}{\sqrt{q}} \qquad \beta_2 = \frac{p^2+6q}{q}$$

$$M(t) = \frac{p}{1-qe^t} \qquad \phi(t) = \frac{p}{1-qe^{it}} \qquad P(t) = \frac{p}{1-qt}$$

# Statistical Methods for Data Analytics

## Hypergeometric Distribution

$$p(x) = \frac{\binom{M}{x}\binom{N-M}{n-x}}{\binom{N}{n}} \quad x = 0,1,2,\ldots,n \quad x \leq M \quad n-x \leq N-M$$

$$n, M, N, \in N \quad 1 \leq n \leq N \quad 1 \leq M \leq N \quad N = 1,2,\ldots$$

$$\mu = n\frac{M}{N} \quad \sigma^2 = \left(\frac{N-n}{N-1}\right)n\frac{M}{N}\left(1-\frac{M}{N}\right) \quad \beta_1 = \frac{(N-2M)(N-2n)\sqrt{N-1}}{(N-2)\sqrt{nM(N-M)(N-n)}}$$

$$\beta_2 = \frac{N^2(N-1)}{(N-2)(N-3)nM(N-M)(N-n)}$$

$$\left\{N(N+1) - 6n(N-n) + 3\frac{M}{N^2}(N-M)\left[N^2(n-2) - Nn^2 + 6n(N-n)\right]\right\}$$

$$M(t) = \frac{(N-M)!(N-n)!}{N!}F(.,e^t) \quad \phi(t) = \frac{(N-M)!(N-n)!}{N!}F(.,e^{it})$$

$$P(t) = \left(\frac{N-M}{N}\right)^n F(.,t)$$

$F(\alpha, \beta, \gamma, x)$ is the hypergeometric function. $\alpha = -n;$ $\beta = -M;$ $\gamma = N - M - n + 1$

## Negative Binomial Distribution

$$p(x) = \binom{x+r-1}{r-1}p^r q^x \quad x = 0,1,2,\ldots \quad r = 1,2,\ldots \quad 0 \leq p \leq 1 \quad q = 1-p$$

$$\mu = \frac{rq}{p} \quad \sigma^2 = \frac{rq}{p^2} \quad \beta_1 = \frac{2-p}{\sqrt{rq}} \quad \beta_2 = 3 + \frac{p^2 + 6q}{rq}$$

$$M(t) = \left(\frac{p}{1-qe^t}\right)^r \quad \phi(t) = \left(\frac{p}{1-qe^{it}}\right)^r \quad P(t) = \left(\frac{p}{1-qt}\right)^r$$

## Poisson Distribution

$$p(x) = \frac{e^{-\mu}\mu^x}{x!} \qquad x = 0,1,2,\ldots \qquad \mu > 0$$

$$\mu = \mu \qquad \sigma^2 = \mu \qquad \beta_1 = \frac{1}{\sqrt{\mu}} \qquad \beta_2 = 3 + \frac{1}{\mu}$$

$$M(t) = \exp\left[\mu(e^t - 1)\right] \qquad \sigma(t) = \exp\left[\mu(e^{it} - 1)\right] \qquad P(t) = \exp\left[\mu(t - 1)\right]$$

## Rectangular (Discrete Uniform) Distribution

$$p(x) = 1/n \qquad x = 1,2,\ldots,n \qquad n \in N$$

$$\mu = \frac{n+1}{2} \qquad \sigma^2 = \frac{n^2-1}{12} \qquad \beta_1 = 0 \qquad \beta_2 = \frac{3}{5}\left(3 - \frac{4}{n^2-1}\right)$$

$$M(t) = \frac{e^t\left(1-e^{nt}\right)}{n\left(1-e^t\right)} \qquad \phi(t) = \frac{e^{it}\left(1-e^{nit}\right)}{n\left(1-e^{it}\right)} \qquad P(t) = \frac{t\left(1-t^n\right)}{n(1-t)}$$

# CONTINUOUS DISTRIBUTIONS

Probability density function, $f(x)$
    Mean, $\mu$
    Variance, $\sigma^2$
    Coefficient of skewness, $\beta_1$
    Coefficient of kurtosis, $\beta_2$
    Moment-generating function, $M(t)$
    Characteristic function, $\phi(t)$

## Arcsin Distribution

$$f(x) = \frac{1}{\pi\sqrt{x(1-x)}} \qquad 0 < x < 1$$

$$\mu = \frac{1}{2} \qquad \sigma^2 = \frac{1}{8} \qquad \beta_1 = 0 \qquad \beta_2\,\frac{3}{2}$$

# Statistical Methods for Data Analytics

## Beta Distribution

$$f(x) = \frac{\Gamma(\alpha+\beta)}{\Gamma(\alpha)\Gamma(\beta)} x^{\alpha-1}(1-x)^{\beta-1} \qquad 0 < x < 1 \qquad \alpha, \beta > 0$$

$$\mu = \frac{\alpha}{\alpha+\beta} \qquad \sigma^2 = \frac{\alpha\beta}{(\alpha+\beta)^2(\alpha+\beta+1)} \qquad \beta_1 = \frac{2(\beta-\alpha)\sqrt{\alpha+\beta+1}}{\sqrt{\alpha\beta}(\alpha+\beta+2)}$$

$$\beta_2 = \frac{3(\alpha+\beta+1)\left[2(\alpha+\beta)^2 + \alpha\beta(\alpha+\beta-6)\right]}{\alpha\beta(\alpha+\beta+2)(\alpha+\beta+3)}$$

## Cauchy Distribution

$$f(x) = \frac{1}{b\pi\left[1 + \left(\frac{x-a}{b}\right)^2\right]} \qquad -\infty < x < \infty \qquad -\infty < a < \infty \qquad b > 0$$

$\mu, \sigma^2, \beta_1, \beta_2, M(t)$ do not exist. $\phi(t) = \exp\left[ait - b|t|\right]$

## Chi Distribution

$$f(x) = \frac{x^{n-1}e^{-x^2/2}}{2^{(n/2)-1}\Gamma(n/2)} \qquad x \geq 0 \qquad n \in N$$

$$\mu = \frac{\Gamma\left(\frac{n+1}{2}\right)}{\Gamma\left(\frac{n}{2}\right)} \qquad \sigma^2 = \frac{\Gamma\left(\frac{n+2}{2}\right)}{\Gamma\left(\frac{n}{2}\right)} - \left[\frac{\Gamma\left(\frac{n+1}{2}\right)}{\Gamma\left(\frac{n}{2}\right)}\right]^2$$

## Chi-Square Distribution

$$f(x) = \frac{e^{-x/2}x^{(v/2)-1}}{2^{v/2}\Gamma(v/2)} \qquad x \geq 0 \qquad v \in N$$

$$\mu = v \qquad \sigma^2 = 2v \qquad \beta_1 = 2\sqrt{2/v} \qquad \beta_2 = 3 + \frac{12}{v} \qquad M(t) = (1-2t)^{-v/2}, \quad t < \frac{1}{2}$$

$$\phi(t) = (1-2it)^{-v/2}$$

## ERLANG DISTRIBUTION

$$f(x) = \frac{1}{\beta^{n(n-1)!}} x^{n-1} e^{-x/\beta} \qquad x \geq 0 \qquad \beta > 0 \qquad n \in N$$

$$\mu = n\beta \qquad \sigma^2 = n\beta^2 \qquad \beta_1 = \frac{2}{\sqrt{n}} \qquad \beta_2 = 3 + \frac{6}{n}$$

$$M(t) = (1 - \beta t)^{-n} \qquad \phi(t) = (1 - \beta it)^{-n}$$

## EXPONENTIAL DISTRIBUTION

$$f(x) = \lambda e^{-\lambda x} \qquad x \geq 0 \qquad \lambda > 0$$

$$\mu = \frac{1}{\lambda} \qquad \sigma^2 = \frac{1}{\lambda^2} \qquad \beta_1 = 2 \qquad \beta_2 = 9 \qquad M(t) = \frac{\lambda}{\lambda - t}$$

$$\phi(t) = \frac{\lambda}{\lambda - it}$$

## EXTREME-VALUE DISTRIBUTION

$$f(x) = \exp\left[-e^{-(x-\alpha)/\beta}\right] \qquad -\infty < x < \infty \qquad -\infty < \alpha < \infty \qquad \beta > 0$$

$$\mu = \alpha + \gamma\beta, \quad \gamma \doteq .5772\ldots \text{ is Euler's constant } \sigma^2 = \frac{\pi^2 \beta^2}{6}.$$

$$\beta_1 = 1.29857 \qquad \beta_2 = 5.4$$

$$M(t) = e^{\alpha t}\Gamma(1-\beta t), \quad t < \frac{1}{\beta} \qquad \phi(t) = e^{\alpha it}\Gamma(1-\beta it)$$

## F DISTRIBUTION

$$f(x) \frac{\Gamma[(v_1 + v_2)/2] v_1^{v_1/2} v_2^{v_2/2}}{\Gamma(v_1/2)\Gamma(v_2/2)} x^{(v_1/2)-1} (v_2 + v_1 x)^{-(v_1+v_2)/2}$$

$$x > 0 \qquad v_1, v_2 \in N$$

$$\mu = \frac{v_2}{v_2 - 2}, v_2 \geq 3 \qquad \sigma^2 = \frac{2v_2^2(v_1 + v_2 - 2)}{v_1(v_2 - 2)^2(v_2 - 4)}, \quad v_2 \geq 5$$

$$\beta_1 = \frac{(2v_1 + v_2 - 2)\sqrt{8(v_2 - 4)}}{\sqrt{v_1}(v_2 - 6)\sqrt{v_1 + v_2 - 2}}, \quad v_2 \geq 7$$

$$\beta_2 = 3 + \frac{12\left[(v_2 - 2)^2(v_2 - 4) + v_1(v_1 + v_2 - 2)(5v_2 - 22)\right]}{v_1(v_2 - 6)(v_2 - 8)(v_1 + v_2 - 2)}, \quad v_2 \geq 9$$

$M(t)$ does not exist. $\phi\left(\dfrac{v_1}{v_2}t\right) = \dfrac{G(v_1, v_2, t)}{B(v_1/2, v_2/2)}$

$B(a,b)$ is the beta function. $G$ is defined by

$$(m + n - 2)G(m,n,t) = (m - 2)G(m - 2, n, t) + 2it\, G(m, n - 2, t), \quad m, n > 2$$

$$mG(m,n,t) = (n - 2)G(m + 2, n - 2, t) - 2it\, G(m + 2, n - 4, t), \quad n > 4$$

$$nG(2, n, t) = 2 + 2it\, G(2, n - 2, t), \quad n > 2$$

## Gamma Distribution

$$f(x) = \frac{1}{\beta^\alpha \Gamma(\alpha)} x^{\alpha - 1} e^{-x/\beta} \qquad x \geq 0 \qquad \alpha, \beta > 0$$

$$\mu = \alpha\beta \qquad \sigma^2 = \alpha\beta^2 \qquad \beta_1 = \frac{2}{\sqrt{\alpha}} \qquad \beta_2 = 3\left(1 + \frac{2}{\alpha}\right)$$

$$M(t) = (1 - \beta t)^{-\alpha} \qquad \phi(t) = (1 - \beta it)^{-\alpha}$$

## Half-Normal Distribution

$$f(x) = \frac{2\theta}{\pi} \exp\left[-\left(\theta^2 x^2 / \pi\right)\right] \qquad x \geq 0 \qquad \theta > 0$$

$$\mu = \frac{1}{\theta} \qquad \sigma^2 = \left(\frac{\pi-2}{2}\right)\frac{1}{\theta^2} \qquad \beta_1 = \frac{4-\pi}{\theta^3} \qquad \beta_2 = \frac{3\pi^2 - 4\pi - 12}{4\theta^4}$$

## LAPLACE (DOUBLE EXPONENTIAL) DISTRIBUTION

$$f(x) = \frac{1}{2\beta}\exp\left[-\frac{|x-\alpha|}{\beta}\right] \qquad -\infty < x < \infty \qquad -\infty < \alpha < \infty \qquad \beta > 0$$

$$\mu = \alpha \qquad \sigma^2 = 2\beta^2 \qquad \beta_1 = 0 \qquad \beta_2 = 6$$

$$M(t) = \frac{e^{\alpha t}}{1-\beta^2 t^2} \qquad \phi(t) = \frac{e^{\alpha it}}{1+\beta^2 t^2}$$

## LOGISTIC DISTRIBUTION

$$f(x) = \frac{\exp[(x-\alpha)/\beta]}{\beta\left(1+\exp[(x-\alpha)/\beta]\right)^2}$$

$$-\infty < x < \infty \qquad -\infty < \alpha < \infty \qquad -\infty < \beta < \infty$$

$$\mu = \alpha \qquad \sigma^2 = \frac{\beta^2 \pi^2}{3} \qquad \beta_1 = 0 \qquad \beta_2 = 4.2$$

$$M(t) = e^{\alpha t}\pi\beta t\,\csc(\pi\beta t) \qquad \phi(t) = e^{\alpha it}\pi\beta it\,\csc(\pi\beta it)$$

## LOGNORMAL DISTRIBUTION

$$f(x) = \frac{1}{\sqrt{2\pi}\sigma x}\exp\left[-\frac{1}{2\sigma^2}(\ln x - \mu)^2\right]$$

$$x > 0 \qquad -\infty < \mu < \infty \qquad \sigma > 0$$

$$\mu = e^{\mu + \sigma^2/2} \qquad \sigma^2 = e^{2\mu+\sigma^2}\left(e^{\sigma^2}-1\right)$$

$$\beta_1 = \left(e^{\sigma^2}+2\right)\left(e^{\sigma^2}-1\right)^{1/2} \qquad \beta_2 = \left(e^{\sigma^2}\right)^4 + 2\left(e^{\sigma^2}\right)^3 + 3\left(e^{\sigma^2}\right)^2 - 3$$

## Noncentral Chi-Square Distribution

$$f(x) = \frac{\exp\left[-\frac{1}{2}(x+\lambda)\right]}{2^{v/2}} \sum_{j=0}^{\infty} \frac{x^{(v/2)+j-1}\lambda^j}{\Gamma\left(\frac{v}{2}+j\right)2^{2j} j!}$$

$$x > 0 \qquad \lambda > 0 \qquad v \in N$$

$$\mu = v + \lambda \qquad \sigma^2 = 2(v + 2\lambda) \qquad \beta_1 = \frac{\sqrt{8}(v + 3\lambda)}{(v + 2\lambda)^{3/2}} \qquad \beta_2 = 3 + \frac{12(v + 4\lambda)}{(v + 2\lambda)^2}$$

$$M(t) = (1 - 2t)^{-v/2} \exp\left[\frac{\lambda t}{1 - 2t}\right] \qquad \phi(t) = (1 - 2it)^{-v/2} \exp\left[\frac{\lambda it}{1 - 2it}\right]$$

## Noncentral F Distribution

$$f(x) = \sum_{i=0}^{\infty} \frac{\Gamma\left(\frac{2i + v_1 + v_2}{2}\right)\left(\frac{v_1}{v_2}\right)^{(2i+v_1)/2} x^{(2i+v_1-2)/2} e^{-\lambda/2} \left(\frac{\lambda}{2}\right)^i}{\Gamma\left(\frac{v_2}{2}\right)\Gamma\left(\frac{2i + v_1}{2}\right) v_1! \left(1 + \frac{v_1}{v_2}x\right)^{(2i+v_1+v_2)/2}}$$

$$x > 0 \qquad v_1, v_2 \in N \qquad \lambda > 0$$

$$\mu = \frac{(v_1 + \lambda)v_2}{(v_2 - 2)v_1}, \qquad v_2 > 2$$

$$\sigma^2 = \frac{(v_1 + \lambda)^2 + 2(v_1 + \lambda)v_2^2}{(v_2 - 2)(v_2 - 4)v_1^2} - \frac{(v_1 + \lambda)^2 v_2^2}{(v_2 - 2)^2 v_1^2}, \qquad v_2 > 4$$

## Noncentral T Distribution

$$f(x) = \frac{v^{v/2}}{\Gamma\left(\frac{v}{2}\right)} \frac{e^{-\delta^2/2}}{\sqrt{\pi}(v + x^2)^{(v+1)/2}} \sum_{i=0}^{\infty} \Gamma\left(\frac{v + i + 1}{2}\right)\left(\frac{\delta^i}{i!}\right)\left(\frac{2x^2}{v + x^2}\right)^{i/2}$$

$$-\infty < x < \infty \qquad -\infty < \delta < \infty \qquad v \in N$$

$$\mu'_r = c_r \frac{\Gamma\left(\frac{v-r}{2}\right) v^{r/2}}{2^{r/2} \Gamma\left(\frac{v}{2}\right)}, \quad v > r, \quad c_{2r-1} = \sum_{i=1}^{r} \frac{(2r-1)! \delta^{2r-1}}{(2i-1)!(r-i)! 2^{r-i}},$$

$$c_{2r} = \sum_{i=0}^{r} \frac{(2r)! \delta^{2i}}{(2i)!(r-i)! 2^{r-i}}, \quad r = 1,2,3,\ldots$$

## Normal Distribution

$$f(x) = \frac{1}{\sigma\sqrt{2\pi}} \exp\left[-\frac{(x-\mu)^2}{2\sigma^2}\right]$$

$$-\infty < x < \infty \quad -\infty < \mu < \infty \quad \sigma > 0$$

$$\mu = \mu \quad \sigma^2 = \sigma^2 \quad \beta_1 = 0 \quad \beta_2 = 3 \quad M(t) = \exp\left[\mu t + \frac{t^2 \sigma^2}{2}\right]$$

$$\phi(t) = \exp\left[\mu i t - \frac{t^2 \sigma^2}{2}\right]$$

## Pareto Distribution

$$f(x) = \theta a^\theta / x^{\theta+1} \quad x \geq a \quad \theta > 0 \quad a > 0$$

$$\mu = \frac{\theta a}{\theta - 1}, \quad \theta > 1 \quad \sigma^2 = \frac{\theta a^2}{(\theta-1)^2 (\theta-2)}, \quad \theta > 2$$

$M(t)$ does not exist.

## Rayleigh Distribution

$$f(x) = \frac{x}{\sigma^2} \exp\left[-\frac{x^2}{2\sigma^2}\right] \quad x \geq 0 \quad \sigma = 0$$

# Statistical Methods for Data Analytics

$$\mu = \sigma\sqrt{\pi/2} \qquad \sigma^2 = 2\sigma^2\left(1-\frac{\pi}{4}\right) \qquad \beta_1 = \frac{\sqrt{\pi}}{4}\frac{(\pi-3)}{\left(1-\frac{\pi}{4}\right)^{3/2}}$$

$$\beta_2 = \frac{2-\frac{3}{16}\pi^2}{\left(1-\frac{\pi}{4}\right)^2}$$

## *T* Distribution

$$f(x) = \frac{1}{\sqrt{\pi v}}\frac{\Gamma\left(\frac{v+1}{2}\right)}{\Gamma\frac{v}{2}}\left(1+\frac{x^2}{v}\right)^{-(v+1)/2} \qquad -\infty < x < \infty \qquad v \in N$$

$$\mu = 0, \quad v \geq 2 \qquad \sigma^2 = \frac{v}{v-2}, \quad v \geq 3 \qquad \beta_1 = 0, \quad v \geq 4$$

$$\beta_2 = 3 + \frac{6}{v-4}, \quad v \geq 5$$

$$M(t) \text{ does not exist. } \phi(t) = \frac{\sqrt{\pi}\Gamma\left(\frac{v}{2}\right)}{\Gamma\left(\frac{v+1}{2}\right)}\int_{-\infty}^{\infty}\frac{e^{itz\sqrt{v}}}{(1+z^2)^{(v+1)/2}}dz$$

## Triangular Distribution

$$f(x) = \begin{cases} 0 & x \leq a \\ 4(x-a)/(b-a)^2 & a < x \leq (a+b)/2 \\ 4(b-x)/(b-a)^2 & (a+b)/2 < x < b \\ 0 & x \geq b \end{cases}$$

$$-\infty < a < b < \infty$$

$$\mu = \frac{a+b}{2} \qquad \sigma^2 = \frac{(b-a)^2}{24} \qquad \beta_1 = 0 \qquad \beta_2 = \frac{12}{5}$$

$$M(t) = -\frac{4\left(e^{at/2}-e^{bt/2}\right)^2}{t^2(b-a)^2} \qquad \phi(t) = \frac{4\left(e^{ait/2}-e^{bit/2}\right)^2}{t^2(b-a)^2}$$

## Uniform Distribution

$$f(x) = \frac{1}{b-a} \qquad a \leq x \leq b \qquad -\infty < a < b < \infty$$

$$\mu = \frac{a+b}{2} \qquad \sigma^2 = \frac{(b-a)^2}{12} \qquad \beta_1 = 0 \qquad \beta_2 = \frac{9}{5}$$

$$M(t) = \frac{e^{bt} - e^{at}}{(b-a)t} \qquad \phi(t) = \frac{e^{bit} - e^{ait}}{(b-a)it}$$

## Weibull Distribution

$$f(x) = \frac{\alpha}{\beta^\alpha} x^{\alpha-1} e^{-(x/\beta)^\alpha} \qquad x \geq 0 \qquad \alpha, \beta > 0$$

$$\mu = \beta\, \Gamma\left(1 + \frac{1}{\alpha}\right) \qquad \sigma^2 = \beta^2\left[\Gamma\left(1 + \frac{2}{\alpha}\right) - \Gamma^2\left(1 + \frac{1}{\alpha}\right)\right]$$

$$\beta_1 = \frac{\Gamma\left(1 + \frac{3}{\alpha}\right) - 3\Gamma\left(1 + \frac{1}{\alpha}\right)\Gamma\left(1 + \frac{2}{\alpha}\right) + 2\Gamma^3\left(1 + \frac{1}{\alpha}\right)}{\left[\Gamma\left(1 + \frac{2}{\alpha}\right) - \Gamma^2\left(1 + \frac{1}{\alpha}\right)\right]^{3/2}}$$

$$\beta_2 = \frac{\Gamma\left(1 + \frac{4}{\alpha}\right) - 4\Gamma\left(1 + \frac{1}{\alpha}\right)\Gamma\left(1 + \frac{3}{\alpha}\right) + 6\Gamma^2\left(1 + \frac{1}{\alpha}\right)\Gamma\left(1 + \frac{2}{\alpha}\right) - 3\Gamma^4\left(1 + \frac{1}{\alpha}\right)}{\left[\Gamma\left(1 + \frac{2}{\alpha}\right) - \Gamma^2\left(1 + \frac{1}{\alpha}\right)\right]^2}$$

## DISTRIBUTION PARAMETERS

### Average

$$\bar{x} = \frac{1}{n}\sum_{i=1}^{n} x_i$$

### Variance

$$s^2 = \frac{1}{n-1}\sum_{i=1}^{n}(x_i - \bar{x})^2$$

## STANDARD DEVIATION

$$s = \sqrt{s^2}$$

## STANDARD ERROR

$$\frac{s}{\sqrt{n}}$$

## SKEWNESS

(missing if $s=0$ or $n<3$)

$$\frac{n\sum_{i=1}^{n}(x_i - \bar{x})^3}{(n-1)(n-2)s^3}$$

## STANDARDIZED SKEWNESS

$$\frac{\text{skewness}}{\sqrt{\frac{6}{n}}}$$

## KURTOSIS

(missing if $s=0$ or $n<4$)

$$\frac{n(n+1)\sum_{i=1}^{n}(x_i - \bar{x})^4}{(n-1)(n-2)(n-3)s^4} - \frac{3(n-1)^2}{(n-2)(n-3)}$$

## STANDARDIZED KURTOSIS

$$\frac{\text{Kurtosis}}{\sqrt{\frac{24}{n}}}$$

## WEIGHTED AVERAGE

$$\frac{\sum_{i=1}^{n} x_i w_i}{\sum_{i=1}^{n} w_i}$$

# ESTIMATION AND TESTING

## $100(1-\alpha)\%$ Confidence Interval for Mean

$$\bar{x} \pm t_{n-1;\alpha/2} \frac{s}{\sqrt{n}}$$

## $100(1-\alpha)\%$ Confidence Interval for Variance

$$\left[ \frac{(n-1)s^2}{\chi^2_{n-1;\alpha/2}}, \frac{(n-1)s^2}{\chi^2_{n-1;1-\alpha/2}} \right]$$

## $100(1-\alpha)\%$ Confidence Interval for Difference in Means

### Equal Variance

$$(\bar{x}_1 - \bar{x}_2) \pm t_{n_1+n_2-2;\alpha/2}\, s_p \sqrt{\frac{1}{n_1} + \frac{1}{n_2}}$$

where

$$s_p = \sqrt{\frac{(n_1-1)s_1^2 + (n_2-1)s_2^2}{n_1+n_2-2}}$$

### Unequal Variance

$$\left[ (\bar{x}_1 - \bar{x}_2) \pm t_{m;\alpha/2} \sqrt{\frac{s_1^2}{n_1} + \frac{s_2^2}{n_2}} \right]$$

where

$$\frac{1}{m} = \frac{c^2}{n_1-1} + \frac{(1-c)^2}{n_2-1}$$

and

$$c = \frac{\dfrac{s_1^2}{n_1}}{\dfrac{s_1^2}{n_1} + \dfrac{s_2^2}{n_2}}$$

## $100(1-\alpha)\%$ Confidence Interval for Ratio of Variances

$$\left(\frac{s_1^2}{s_2^2}\right)\left(\frac{1}{F_{n_1-1,n_2-1;\,\alpha/2}}\right), \left(\frac{s_1^2}{s_2^2}\right)\left(\frac{1}{F_{n_1-1,n_2-1;\,\alpha/2}}\right)$$

## Normal Probability Plot

The data is sorted from the smallest to the largest value to compute order statistics. A scatterplot is then generated where

$$\text{Horizontal position} = x_{(i)}$$

$$\text{Vertical position} = \Phi\left(\frac{i-3/8}{n+1/4}\right)$$

The labels for the vertical axis are based upon the probability scale using

$$100\left(\frac{i-3/8}{n+1/4}\right)$$

## Comparison of Poisson Rates

$$n_j = \text{number of events in sample } j$$

$$t_j = \text{length of sample } j$$

$$\text{Rate estimates: } r_j = \frac{n_j}{t_j}$$

$$\text{Rate ratio: } \frac{r_1}{r_2}$$

Test statistic:

$$z = \max\left(0, \frac{\left|n_1 - \frac{(n_1+n_2)}{2}\right| - \frac{1}{2}}{\sqrt{\frac{(n_1+n_2)}{4}}}\right)$$

where $z$ follows the standard normal distribution.

# DISTRIBUTION FUNCTIONS AND PARAMETER ESTIMATION

## BERNOULLI

$$\hat{p} = \bar{x}$$

## BINOMIAL

$$\hat{p} = \frac{\bar{x}}{n}$$

where $n$ is the number of trials.

## DISCRETE UNIFORM

$$\hat{a} = \min x_i$$

$$\hat{b} = \max x_i$$

## GEOMETRIC

$$\hat{p} = \frac{1}{1+\bar{x}}$$

## NEGATIVE BINOMIAL

$$\hat{p} = \frac{k}{\bar{x}}$$

where $k$ is the number of successes.

## POISSON

$$\hat{\beta} = \bar{x}$$

## BETA

$$\hat{\alpha} = \bar{x}\left[\frac{\bar{x}(1-\bar{x})}{s^2} - 1\right]$$

$$\hat{\beta} = (1-\bar{x})\left(\frac{\bar{x}(1-\bar{x})}{s^2} - 1\right)$$

# Statistical Methods for Data Analytics

## CHI-SQUARE

$$\text{d.f. } \bar{v} = \bar{x}$$

## ERLANG

$$\hat{\alpha} = \text{round}\left(\hat{\alpha} \text{ from gamma}\right)$$

$$\hat{\beta} = \frac{\hat{\alpha}}{\bar{x}}$$

## EXPONENTIAL

$$\hat{\beta} = \frac{1}{\bar{x}}$$

Note: system displays $1/\hat{\beta}$.

## F

$$\text{Numerator degrees of freedom (num d f): } \hat{v} = \frac{2\hat{w}^3 - 4\hat{w}^2}{\left(s^2(\hat{w}-2)^2(\hat{w}-4)\right) - 2\hat{w}^2}$$

$$\text{Denominator degrees of freedom (den df): } \hat{w} = \frac{\max(1, 2\bar{x})}{-1 + \bar{x}}$$

## GAMMA

$$R = \log\left(\frac{\text{arithmetic mean}}{\text{geometric mean}}\right)$$

If $0 < R \leq 0.5772$,

$$\hat{\alpha} = R^{-1}\left(0.5000876 + 0.1648852\, R - 0.0544274\, R\right)^2$$

or if $R > 0.5772$,

$$\hat{\alpha} = R^{-1}\left(17.79728 + 11.968477\, R + R^2\right)^{-1}\left(8.898919 + 9.059950\, R + 0.9775373\, R^2\right)$$

$$\hat{\beta} = \hat{\alpha} / \bar{x}$$

## LOGNORMAL

$$\hat{\mu} = \frac{1}{n}\sum_{i=1}^{n} \log x_i$$

$$\hat{\alpha} = \sqrt{\frac{1}{n-1}\sum_{i=1}^{n}(\log x_i - \hat{\mu})^2}$$

### SYSTEM DISPLAYS

Means: $\exp(\hat{\mu} + \hat{\alpha}^2/2)$

Standard deviation: $\sqrt{\exp(2\hat{\mu} + \hat{\alpha}^2)[\exp(\hat{\alpha}^2)-1]}$

## NORMAL

$$\hat{\mu} = \bar{x}$$

$$\hat{\sigma} = s$$

## STUDENT'S T

If $s^2 \leq 1$ or if $\hat{v} \leq 2$, then the system indicates that the data is inappropriate.

$$s^2 = \frac{\sum_{i=1}^{n} x_i^2}{n}$$

$$\hat{v} = \frac{2s^2}{-1+s^2}$$

## TRIANGULAR

$$\hat{a} = \min x_i$$

$$\hat{c} = \max x_i$$

$$\hat{b} = 3\bar{x} - \hat{a} - \bar{x}$$

## Uniform

$$\hat{a} = \min x_i$$

$$\hat{b} = \max x_i$$

## Weibull

Solves the simultaneous equations:

$$\hat{\alpha} = \dfrac{n}{\left[\left(\dfrac{1}{\hat{\beta}}\right)\sum_{i=1}^{n} x_i^{\hat{\alpha}} \log x_i - \sum_{i=1}^{n} \log x_i\right]}$$

$$\hat{\beta} = \left(\dfrac{\sum_{i=1}^{n} x_i^{\hat{\alpha}}}{n}\right)^{\frac{1}{\hat{\alpha}}}$$

## Chi-Square Test for Distribution Fitting

Divide the range of data into non-overlapping classes. The classes are aggregated at each end to ensure that classes have an expected frequency of at least 5.

$O_i$ = observed frequency in class $i$
$E_i$ = expected frequency in class $i$ from fitted distribution
$k$ = number of classes after aggregation

Test statistic:

$$\chi^2 = \sum_{i=1}^{k} \dfrac{(O_i - E_i)^2}{E_i}$$

This statistic follows a chi-square distribution with the degrees of freedom equal to ($k - 1$ – number of estimated parameters)

## Kolmogorov–Smirnov Test

$$D_n^+ = \max\left\{\dfrac{i}{n} - \hat{F}(x_i)\right\}$$

$$1 \leq i \leq n$$

$$D_n^- = \max\left\{\hat{F}(x_i) - \frac{i-1}{n}\right\}$$

$$1 \leq i \leq n$$

$$D_n = \max\{D_n^+, D_n^-\}$$

where $\hat{F}(x_i)$ is the estimated cumulative distribution at $x_i$.

## ANOVA (Analysis of Variance)

### Notations

$$k = \text{number of treatments}$$

$$n_t = \text{number of observations for treatment } t$$

$$\bar{n} = n/k = \text{average treatment size}$$

$$n = \sum_{t=1}^{k} n_t$$

$$x_{it} = i^{\text{th}} \text{ observation in treatment } t$$

$$\bar{x}_t = \text{treatment mean} = \frac{\sum_{i=1}^{n_t} x_{it}}{n_t}$$

$$s_t^2 = \text{treatment variance} = \frac{\sum_{i=1}^{n_t}(x_{it} - \bar{x}_t)^2}{n_t - 1}$$

$$\text{MSE} = \text{mean square error} = \frac{\sum_{t=1}^{k}(n_t - 1)s_t^2}{\left(\sum_{t=1}^{k} n_t\right) - k}$$

$$\text{df} = \text{degrees of freedom for the error term} = \left(\sum_{t=1}^{k} n_t\right) - k$$

## Standard Error (Internal)

$$\sqrt{\frac{s_t^2}{n_t}}$$

## Standard Error (Pooled)

$$\sqrt{\frac{\text{MSE}}{n_t}}$$

## Interval Estimates

$$\bar{x}_t \pm M\sqrt{\frac{\text{MSE}}{n_t}}$$

where
　Confidence interval

$$M = t_{n-k;\alpha/2}$$

　Least significant difference (LSD) interval

$$M = \frac{1}{\sqrt{2}} t_{n-k;\alpha/2}$$

## Tukey Interval

$$M = \frac{1}{2} q_{n-k,k;\alpha}$$

where $q_{n-k,k;\alpha}$ is the value of the studentized range distribution with $n-k$ degrees of freedom and $k$ samples such that the cumulative probability equals $1-\alpha$.

## Scheffe Interval

$$M = \frac{\sqrt{k-1}}{\sqrt{2}} \sqrt{F_{k-1,\,n-k;\,\alpha}}$$

## Cochran C-Test

Follow $F$ distribution with $\bar{n}-1$ and $(\bar{n}-1)(k-1)$ degrees of freedom.

$$\text{Test statistic:} \quad F = \frac{(k-1)C}{1-C}$$

where

$$C = \frac{\max s_t^2}{\sum_{t=1}^{k} s_t^2}$$

## BARTLETT TEST

Test statistic:

$$B = 10^{\frac{M}{(n-k)}}$$

$$M = (n-k)\log_{10} \text{MSE} - \sum_{t=1}^{k}(n_t - 1)\log_{10} s_t^2$$

The significance test is based on

$$\frac{M(\ln 10)}{1 + \frac{1}{3(k-1)}\left[\sum_{t=1}^{k}\frac{1}{(n_t - 1)} - \frac{1}{N-k}\right]} \chi^2_{k-1}$$

which follows a chi-square distribution with $k - 1$ degrees of freedom.

## HARTLEY'S TEST

$$H = \frac{\max(s_t^2)}{\min(s_t^2)}$$

## KRUSKAL–WALLIS TEST

Average rank of treatment:

$$\bar{R}_t = \frac{\sum_{i=1}^{n_t} R_{it}}{n_t}$$

If there are no ties:

Test statistic: $w = \left(\frac{12}{n}\sum_{i=1}^{k} n_t \bar{R}_t^2\right) - 3(n+1)$

Adjustment for ties

# Statistical Methods for Data Analytics

Let $u_j$ be the number of observations tied at any rank for $j = 1, 2, 3, \ldots, m$ where $m$ is the number of unique values in the sample.

$$W = \frac{w}{1 - \dfrac{\sum_{j=1}^{m} u_j^3 - \sum_{j=1}^{m} u_j}{n(n^2 - 1)}}$$

Significance level: $W$ follows a chi-square distribution with $k - 1$ degrees of freedom.

## FREIDMAN TEST

$$X_{it} = \text{observation in the } i^{th} \text{row}, t^{th} \text{ column}$$

$$i = 1, 2, \ldots, n \quad t = 1, 2, \ldots, k$$

$$R_{it} = \text{rank of } X_{it} \text{ within its row}$$

$n$ = common treatment size (all treatment sizes must be the same for this test)

$$R_t = \sum_{i=1}^{n} R_{it}$$

average rank $\overline{R}_t = \dfrac{\sum_{i=1}^{n_t} R_{it}}{n_t}$

where data are ranked within each row separately.
Test statistic:

$$Q = \frac{12S(k-1)}{nk(k^2 - 1) - \left(\sum u^3 - \sum u\right)}$$

where

$$S = \left(\sum_{t=1}^{k} R_i^2\right) - \frac{n^2 k(k+1)^2}{4}$$

$Q$ follows a chi-square distribution with $k$ degrees of freedom.

## REGRESSION

### Notations

$Y$ = vector of $n$ observations for the dependent variable ~

$X$ = $n$ by $p$ matrix of observations for $p$ independent variables, including constant term, if any

~ indicates that a variable is a vector or matrix.

$$\bar{Y} = \frac{\sum_{i=1}^{n} Y_i}{n}$$

### Regression Statistics

1. Estimated coefficients
   Note: estimated by a modified Gram-Schmidt orthogonal decomposition with tolerance $= 1.0E - 08$.

$$b = (X'X)^{-1} XY$$

2. Standard errors

$$S(b) = \sqrt{\text{diagonal elements of } (X'X)^{-1} \text{MSE}}$$

where $\text{MSE} = \dfrac{Y'Y - b'X'Y}{n - p}$

3. $t$-Values

$$t = \frac{b}{S(b)}$$

4. Significance level
   $t$-Values follow the student's $t$ distribution with $n$-$p$ degrees of freedom.

5. $R$-squared

$$R^2 = \frac{\text{SSTO} - \text{SSE}}{\text{SSTO}},$$

where $\text{SSTO} = \begin{cases} Y' - n\bar{Y}^2, & \text{if constant} \\ YY, & \text{if no constant} \end{cases}$

SSE = residual sum of squares

# Statistical Methods for Data Analytics

6. Adjusted $R$-squared

$$1 - \left(\frac{n-1}{n-p}\right)(1 - R^2)$$

7. Standard error of estimate

$$SE = \sqrt{MSE}$$

8. Predicted values

$$\hat{\underline{Y}} = \underline{X}\underline{b}$$

9. Residuals

$$\underline{e} = \underline{Y} - \hat{\underline{Y}}$$

10. Durbin–Watson statistic

$$D = \frac{\sum_{i=1}^{n-1}(e_{i+1} - e_i)^2}{\sum_{i=1}^{n} e_i^2}$$

11. Mean absolute error

$$\frac{\left(\sum_{i=1}^{n}|e_i|\right)}{n}$$

## Predictions

$\underline{X}_h = m$ by $p$ matrix of independent variables for $m$ predictions

1. Predicted value

$$\hat{\underline{Y}}_h = \underline{X}_h \underline{b}$$

2. Standard error of predictions

$$S(\hat{\underline{Y}}_{h(\text{new})}) = \sqrt{\text{diagonal elements of MSE}\left(1 + \underline{X}_h (\underline{X}'\underline{X})^{-1} \underline{X}'_h\right)}$$

3. Standard error of mean response

$$S(\hat{\underline{Y}}_h) = \sqrt{\text{diagonal elements of MSE}\left(\underline{X}_h (\underline{X}'\underline{X})^{-1} \hat{\underline{X}}_h\right)}$$

4. Prediction matrix results
   Column 1 = index numbers of forecasts
   $2 = \hat{\underset{\sim}{Y}}_h$
   $3 = S(\hat{\underset{\sim}{Y}}_{h(new)})$
   $4 = (\hat{\underset{\sim}{Y}}_h - t_{n-p,\alpha/2}\, S(\hat{\underset{\sim}{Y}}_{h(new)}))$
   $5 = (\hat{\underset{\sim}{Y}}_h + t_{n-p,\alpha/2}\, S(\hat{\underset{\sim}{Y}}_{h(new)}))$
   $6 = \hat{\underset{\sim}{Y}}_h - t_{n-p,\alpha/2}\, S(\hat{\underset{\sim}{Y}}_h)$
   $7 = \hat{\underset{\sim}{Y}}_h + t_{n-p,\alpha/2}\, S(\hat{\underset{\sim}{Y}}_h)$

**Nonlinear Regression**

$F(X, \hat{\beta})$ are values of nonlinear function using parameter estimates $\hat{\beta}$.

1. Estimated coefficients
   Obtained by minimizing the residual sum of squares using a search procedure suggested by Marquardt. This is a compromise between Gauss–Newton and steepest descent methods. The user specifies
   a. initial estimates $\beta_0$.
   b. the initial value of Marquardt parameter $\lambda$, which is modified at each iteration.
      As $\lambda \to 0$, the procedure approaches Gauss–Newton; as $\lambda \to \infty$, the procedure approaches steepest descent.
   c. the scaling factor used to multiply Marquardt parameter after each Iteration.
   d. the maximum value of Marquardt parameter.
      Partial derivatives of $F$ with respect to each parameter are estimated numerically.
2. Standard errors
   estimated from residual sum of squares and partial derivatives
3. Ratio

$$\text{ratio} = \frac{\text{coefficient}}{\text{standard error}}$$

4. R-squared

$$R^2 = \frac{\text{SSTO} - \text{SSE}}{\text{SSTO}}$$

where

SSTO $= Y'\underset{\sim}{Y} - n\bar{Y}^2$
SSE = residual sum of squares

# Ridge Regression

Additional notation:

$\underset{\sim}{Z}$ = matrix of independent variables standardized so that $\underset{\sim}{Z}'\underset{\sim}{Z}$ equals the correlation matrix
$\theta$ = value of the ridge parameter

Parameter estimates

$$\underset{\sim}{b}(\theta) = \left(\underset{\sim}{Z}'\underset{\sim}{Z} + \theta I_p\right)^{-1} \underset{\sim}{Z}'\underset{\sim}{Y},$$

# Quality Control

For all quality control formulas:

$$k = \text{number of subgroups}$$

$$n_j = \text{number of observations in subgroup } j$$
$$j = 1, 2, \ldots, k$$

$$x_{ij} = i^{\text{th}} \text{ observation in subgroup } j$$

All formulas below for quality control assume 3-sigma limits. If other limits are specified, the formulas are adjusted proportionally based on sigma for the selected limits. Also, average sample size is used unless otherwise specified.

## Subgroup Statistics

*Subgroup Means*

$$\bar{x}_j = \frac{\sum_{i=1}^{n_j} x_{ij}}{n_j}$$

*Subgroup Standard Deviations*

$$s_j = \sqrt{\frac{\sum_{i=1}^{n_j} (x_{ij} - \bar{x}_j)^2}{(n_j - 1)}}$$

*Subgroup Range*

$$R_j = \max\{x_{ij} | 1 \leq i \leq n_j\} - \min\{x_{ij} | 1 \leq i \leq n_j\}$$

**X-Bar Charts**

$$\text{Compute } \bar{\bar{x}} = \frac{\sum_{j=1}^{k} n_i \bar{x}_j}{\sum_{j=1}^{k} n_i}$$

$$\bar{R} = \frac{\left(\sum_{j=1}^{k} n_i R_j\right)}{\sum_{j=1}^{k} n_i}$$

$$s_p = \sqrt{\frac{\sum_{j=1}^{k}(n_j - 1)s_j^2}{\sum_{j=1}^{k}(n_j - 1)}}$$

$$\bar{n} = \frac{1}{k}\sum_{j=1}^{k} n_i$$

For a chart based on range:

$$\text{Upper Control Limit: UCL} = \bar{\bar{x}} + A_2 \bar{R}$$

$$\text{Lower Control Limit: LCL} = \bar{\bar{x}} - A_2 \bar{R}$$

For a chart based on sigma:

$$\text{UCL} = \bar{\bar{x}} + \frac{3s_p}{\sqrt{\bar{n}}}$$

$$\text{LCL} = \bar{\bar{x}} - \frac{3s_p}{\sqrt{\bar{n}}}$$

# Statistical Methods for Data Analytics

For a chart based on known sigma:

$$\text{UCL} = \bar{\bar{x}} + 3\frac{\sigma}{\sqrt{n}}$$

$$\text{LCL} = \bar{\bar{x}} - 3\frac{\sigma}{\sqrt{n}}$$

If other than 3-sigma limits are used, such as 2-sigma limits, all bounds are adjusted proportionately. If average sample size is not used, then uneven bounds are displays based on

$$1/\sqrt{n_j}$$

rather than $1/\sqrt{n}$.

If the data is normalized, each observation is transformed according to

$$z_{ij} = \frac{x_{ij} - \bar{\bar{x}}}{\hat{\alpha}}$$

where $\hat{\alpha}$ = estimated standard deviation.

## Capability Ratios

Note: The following indices are useful only when the control limits are placed at the specification limits. To override the normal calculations, specify a subgroup size of one and select the "known standard deviation" option. Then enter the standard deviation as half of the distance between the USL and the LSL. Change the position of the centerline to be the midpoint of the USL and LSL and specify the upper and lower control line at one sigma.

$$C_P = \frac{\text{USL} - \text{LSL}}{6\hat{\alpha}}$$

$$C_R = \frac{1}{C_P}$$

$$C_{PK} = \min\left(\frac{\text{USL} - \bar{\bar{x}}}{3\hat{\alpha}}, \frac{\bar{\bar{x}} - \text{LSL}}{3\hat{\alpha}}\right)$$

## *R* Charts

Control Limit: $\text{CL} = \bar{R}$

$$\text{UCL} = D_4 \bar{R}$$

$$LCL = \text{Max}(0, D_3 \bar{R})$$

## S Charts

$$CL = s_P$$

$$UCL = s_P \sqrt{\frac{\chi^2_{n-1;\alpha}}{n-1}}$$

$$LCL = s_P \sqrt{\frac{\chi^2_{n-1;\alpha}}{n-1}}$$

## C Charts

$$\bar{c} = \frac{\sum u_j}{\sum n_j} \qquad \begin{aligned} UCL &= \bar{c} + 3\sqrt{\bar{c}} \\ LCL &= \bar{c} - 3\sqrt{\bar{c}} \end{aligned}$$

where $u_j$ = number of defects in the $j^{th}$ sample.

## U Charts

$$\bar{u} = \frac{\text{number of defects in all samples}}{\text{number of units in all samples}} = \frac{\sum u_j}{\sum n_j}$$

$$UCL = \bar{u} + \frac{3\sqrt{\bar{u}}}{\sqrt{n}}$$

$$LCL = \bar{u} - \frac{3\sqrt{\bar{u}}}{\sqrt{n}}$$

## P Charts

$$p = \frac{\text{number of defective units}}{\text{number of units inspected}}$$

$$\bar{p} = \frac{\text{number of defectives in all samples}}{\text{number of units in all samples}} = \frac{\sum p_j n_j}{\sum n_j}$$

# Statistical Methods for Data Analytics

$$UCL = \bar{p} + \frac{3\sqrt{\bar{p}(1-\bar{p})}}{\sqrt{n}}$$

$$LCL = \bar{p} - \frac{3\sqrt{\bar{p}(1-\bar{p})}}{\sqrt{n}}$$

## NP Charts

$$\bar{p} = \frac{\sum d_j}{\sum n_j},$$

where $d_j$ is the number of defectives in the $j^{th}$ sample.

$$UCL = \bar{n}\,\bar{p} + 3\sqrt{\bar{n}\,\bar{p}(1-\bar{p})}$$

$$LCL = \bar{n}\,\bar{p} - 3\sqrt{\bar{n}\,\bar{p}(1-\bar{p})}$$

## CuSum Chart for the Mean

Control mean $= \mu$
Standard deviation $= \alpha$
Difference to detect $= \Delta$

Plot cumulative sums $C_t$ versus $t$ where

$$C_t = \sum_{i=1}^{t}(\bar{x}_i - \mu) \quad \text{for } t = 1, 2, \ldots, n$$

The V-mask is located at distance

$$d = \frac{2}{\Delta}\left[\frac{\alpha^2/\bar{n}}{\Delta} \ln \frac{1-\beta}{\alpha/2}\right]$$

in front of the last data point.

Angle of mast $= 2\tan^{-1}\dfrac{\Delta}{2}$

Slope of the lines $= \pm\dfrac{\Delta}{2}$

## Multivariate Control Charts

$\underset{\sim}{X}$ = matrix of $n$ rows and $k$ columns containing $n$ observations for each $k$ variable
$S$ = sample covariance matrix
$\underset{\sim}{X_t}$ = observation vector at time $t$
$\underset{\sim}{\bar{X}}$ = vector of column average

Then,

$$T_t^2 = \left(\underset{\sim}{X_t} - \underset{\sim}{\bar{X}}\right) S^{-1} \left(\underset{\sim}{X_t} - \underset{\sim}{\bar{X}}\right)$$

$$\text{UCL} = \left(\frac{k(n-1)}{n-k}\right) F_{k, n-k; \alpha}$$

## Time Series Analysis

### Notations

$$x_t \text{ or } y_t = \text{observation at time } t, t = 1, 2, \ldots, n$$

$$n = \text{number of observations}$$

### Autocorrelation at Lag $k$

$$r_k = \frac{c_k}{c_0}$$

where

$$c_k = \frac{1}{n} \sum_{t=1}^{n-k} (y_t - \bar{y})(y_{t+k} - \bar{y})$$

and

$$\bar{y} = \frac{\left(\sum_{t=1}^{n} y_t\right)}{n}$$

$$\text{standard error} = \sqrt{\frac{1}{n}\left\{1 + 2\sum_{v=1}^{k-1} r_v^2\right\}}$$

## Partial Autocorrelation at Lag *k*

$\hat{\theta}_{kk}$ is obtained by solving the Yule–Walker equations:

$$r_j = \hat{\theta}_{k1} r_{j-1} + \hat{\theta}_{k2} r_{j-2} + \cdots + \hat{\theta}_{k(k-1)} r_{j-k+1} + \hat{\theta}_{kk} r_{j-k}$$

$$j = 1, 2, \ldots, k$$

$$\text{standard error} = \sqrt{\frac{1}{n}}$$

## Cross-Correlation at Lag *k*

*x* = input time series
*y* = output time series

$$r_{xy}(k) = \frac{c_{xy}(k)}{s_x s_y} \quad k = 0, \pm 1, \pm 2, \ldots$$

where

$$c_{xy}(k) = \begin{cases} \dfrac{1}{n} \sum_{t=1}^{n-k} (x_t - \bar{x})(y_{t+k} - \bar{y}) & k = 0, 1, 2, \ldots \\ \dfrac{1}{n} \sum_{t=1}^{n+k} (x_t - \bar{x})(y_{t-k} - \bar{y}) & k = 0, -1, -2, \ldots \end{cases}$$

and

$$S_x = \sqrt{c_{xx}(0)}$$

$$S_y = \sqrt{c_{yy}(0)}$$

## Box-Cox

$$yt = \frac{(y + \lambda_2)^{\lambda_1} - 1}{\lambda_1 g^{(\lambda_1 - 1)}} \quad \text{if } \lambda_1 > 0$$

$$yt = g \ln(y + \lambda_2) \quad \text{if } \lambda_1 = 0$$

where *g* = sample geometric mean $(y + \lambda_2)$.

## Periodogram (Computed Using Fast Fourier Transform)

If $n$ is odd:

$$I(f_1) = \frac{n}{2}\left(a_i^2 + b_i^2\right) \qquad i = 1, 2, \ldots, \left[\frac{n-1}{2}\right]$$

where

$$a_i = \frac{2}{n}\sum_{t=1}^{n} t_t \cos 2\pi f_i t$$

$$b_i = \frac{2}{n}\sum_{t=1}^{n} y_t \sin 2\pi f_i t$$

$$f_i = \frac{i}{n}$$

If $n$ is even, an additional term is added:

$$I(0.5) = n\left(\frac{1}{n}\sum_{t=1}^{n}(-1)^t Y_t\right)^2$$

## CATEGORICAL ANALYSIS

### Notations

$r$ = number of rows in the table
$c$ = number of columns in the table
$f_{ij}$ = frequency in position (row $i$, column $j$)
$x_i$ = distinct values of row variable arranged in ascending order; $i = 1, \ldots, r$
$y_j$ = distinct values of column variable arranged in ascending order; $j = 1, \ldots, c$

### Totals

$$R_j = \sum_{j=1}^{c} f_{ij} \qquad C_j = \sum_{i=1}^{r} f_{ij}$$

$$N = \sum_{i=1}^{r}\sum_{j=1}^{c} f_{ij}$$

Note: any row or column which totals zero is eliminated from the table before calculations are performed.

## Chi-Square

$$\chi^2 = \sum_{i=1}^{r}\sum_{j=1}^{c} \frac{(f_{ij} - E_{ij})^2}{E_{ij}}$$

where

$$E_{ij} = \frac{R_i C_j}{N} \sim \chi^2_{(r-1)(c-1)}$$

A warning is issued if any $E_{ij} < 2$ or if 20% or more of all $E_{ij} < 5$. For $2 \times 2$ tables, a second statistic is printed using Yate's continuity correction.

## Fisher's Exact Test

Run for a $2 \times 2$ table, when $N$ is less than or equal to 100.

## Lambda

$$\lambda = \frac{\left(\sum_{j=1}^{c} f_{max,j} - R_{max}\right)}{N - R_{max}} \text{ with rows dependent}$$

$$\lambda = \frac{\left(\sum_{i=1}^{r} f_{i,max} - C_{max}\right)}{N - C_{max}} \text{ with columns dependent}$$

$$\lambda = \frac{\left(\sum_{i=1}^{r} f_{i,max} + \sum_{j=1}^{c} f_{max,j} - C_{max} - R_{max}\right)}{(2N - R_{max} - C_{max})} \text{ when symmetric}$$

where
 $f_{i\,max}$ = largest value in row $i$
 $f_{max\,j}$ = largest value in column $j$
 $R_{max}$ = largest row total
 $C_{max}$ = largest column total

## Uncertainty Coefficient

$$U_R = \frac{U(R) + U(C) - U(RC)}{U(R)} \text{ with rows dependent}$$

$$U_C = \frac{U(R)+U(C)-U(RC)}{U(C)} \text{ with columns dependent}$$

$$U = 2\left(\frac{U(R)+U(C)-U(RC)}{U(R)+U(C)}\right) \text{ when symmetric}$$

where

$$U(R) = -\sum_{i=1}^{r} \frac{R_i}{N} \log \frac{R_i}{N}$$

$$U(C) = -\sum_{j=1}^{c} \frac{C_j}{N} \log \frac{C_j}{N}$$

$$U(RC) = -\sum_{i=1}^{r}\sum_{j=1}^{c} \frac{f_{ij}}{N} \log \frac{f_{ij}}{N} \qquad \text{for } f_{ij} > 0$$

**Somer's D**

$$D_R = \frac{2(P_C - P_D)}{\left(N^2 - \sum_{j=1}^{c} C_j^2\right)} \text{ with rows dependent}$$

$$D_C = \frac{2(P_C - P_D)}{\left(N^2 - \sum_{i=1}^{r} R_i^2\right)} \text{ with columns dependent}$$

$$D = \frac{4(P_C - P_D)}{\left(N^2 - \sum_{i=1}^{r} R_i^2\right) + \left(N^2 - \sum_{j=1}^{c} C_j^2\right)} \text{ when symmetric}$$

where the number of concordant pairs is

$$P_C = \sum_{i=1}^{r}\sum_{j=1}^{c} f_{ij} \sum_{h<i}\sum_{k<j} f_{hk}$$

and the number of discordant pairs is

$$P_D = \sum_{i=1}^{r}\sum_{j=1}^{c} f_{ij} \sum_{h<i}\sum_{k>j} f_{hk}$$

## Eta

$$E_R = \sqrt{1 - \frac{SS_{RN}}{SS_R}} \text{ with rows dependent}$$

where the total corrected sum of squares for the rows is

$$SS_R = \sum_{i=1}^{r}\sum_{j=1}^{c} x_i^2 f_{ij} - \frac{\left(\sum_{i=1}^{r}\sum_{j=1}^{c} x_i f_{ij}\right)^2}{N}$$

and the sum of squares of rows within categories of columns is

$$SS_{RN} = \sum_{j=1}^{c} \left( \sum_{i=1}^{r} x_i^2 f_{ij} - \frac{\left(\sum_{i=1}^{r} x_i^2 f_{ij}\right)^2}{C_j} \right)$$

$$E_C = \sqrt{1 - \frac{SS_{CN}}{SS_C}} \text{ with columns dependent}$$

where the total corrected sum of squares for the columns is

$$SS_C = \sum_{i=1}^{r}\sum_{j=1}^{c} y_i^2 f_{ij} - \frac{\left(\sum_{i=1}^{r}\sum_{j=1}^{c} y_i f_{ij}\right)^2}{N}$$

and the sum of squares of columns within categories of rows is

$$SS_{CN} = \sum_{i=1}^{r} \left( \sum_{j=1}^{c} y_i^2 f_{ij} - \frac{\left(\sum_{j=1}^{c} y_j^2 f_{ij}\right)^2}{R_i} \right) j$$

## Contingency Coefficient

$$C = \sqrt{\frac{\chi^2}{(\chi^2 + N)}}$$

**Cramer's V**

$$V = \sqrt{\frac{\chi^2}{N}} \text{ for } 2\times 2 \text{ tables}$$

$$V = \sqrt{\frac{\chi^2}{N(m-1)}} \text{ for all others where } m = \min(r, c)$$

**Conditional Gamma**

$$G = \frac{P_C - P_D}{P_C + P_D}$$

**Pearson's r**

$$R = \frac{\sum_{j=1}^{c}\sum_{i=1}^{r} x_i y_j f_{ij} - \frac{\left(\sum_{j=1}^{c}\sum_{i=1}^{r} x_i f_{ij}\right)\left(\sum_{j=1}^{c}\sum_{i=1}^{r} y_i f_{ij}\right)}{N}}{\sqrt{SS_R SS_C}}$$

If $R = 1$, no significance is printed. Otherwise, the one-sided significance is based on

$$t = R\sqrt{\frac{N-2}{1-R^2}}$$

**Kendall's Tau b**

$$\tau = \frac{2(P_C - P_D)}{\sqrt{\left(N^2 - \sum_{i=1}^{r} R_i^2\right)\left(N^2 - \sum_{j=1}^{c} C_j^2\right)}}$$

**Tau C**

$$\tau_C = \frac{2m(P_C - P_D)}{(m-1)N^2}$$

## PROBABILITY TERMINOLOGY

**Experiment.** An activity or occurrence with an observable result
**Outcome.** The result of the experiment
**Sample point.** An outcome of an experiment
**Event.** A set of outcomes (a subset of the sample space) to which a probability is assigned

## Basic Probability Principles

Consider a random sampling process in which all the outcomes solely depend on chance, that is, each outcome is equally likely to happen. If $S$ is a uniform sample space and the collection of desired outcomes is $E$, the probability of the desired outcomes is

$$P(E) = \frac{n(E)}{n(S)}$$

where
$n(E)$ = number of favorable outcomes in $E$
$n(S)$ = number of possible outcomes in $S$

Since $E$ is a subset of $S$,

$$0 \leq n(E) \leq n(S),$$

the probability of the desired outcome is

$$0 \leq P(E) \leq 1$$

## Random Variable

A random variable is a rule that assigns a number to each outcome of a chance experiment.
   Example:

1. A coin is tossed six times. The random variable $X$ is the number of tails that are noted. $X$ can only take the values 1, 2,..., 6, so $X$ is a discrete random variable.
2. A light bulb is burned until it burns out. The random variable $Y$ is its lifetime in hours. $Y$ can take any positive real value, so $Y$ is a continuous random variable.

## Mean Value $\hat{x}$ or Expected Value $\mu$

The mean value or expected value of a random variable indicates its average or central value. It is a useful summary value of the variable's distribution.

1. If random variable $X$ is a discrete mean value,

$$\hat{x} = x_1 p_1 + x_2 p_2 + \ldots + x_n p_n = \sum_{i=1}^{n} x_1 p_1$$

where
$p_i$ = probability densities

2. If $X$ is a continuous random variable with probability density function $f(x)$, then the expected value of $X$ is

$$\mu = E(X) = \int_{-\infty}^{+\infty} xf(x)dx$$

where
$f(x)$ = probability densities

## DISCRETE DISTRIBUTION FORMULAS

Probability mass function, $p(x)$
    Mean, $\mu$
    Variance, $\sigma^2$
    Coefficient of skewness, $\beta_1$
    Coefficient of kurtosis, $\beta_2$
    Moment-generating function, $M(t)$
    Characteristic function, $\phi(t)$
    Probability-generating function, $P(t)$

### BERNOULLI DISTRIBUTION

$$p(x) = p^x q^{x-1} \qquad x = 0,1 \qquad 0 \le p \le 1 \qquad q = 1 - p$$

$$\mu = p \qquad \sigma^2 = pq \qquad \beta_1 = \frac{1-2p}{\sqrt{pq}} \qquad \beta_2 = 3 + \frac{1-6pq}{pq}$$

$$M(t) = q + pe^t \qquad \phi(t) = q + pe^{it} \qquad P(t) = q + pt$$

### BETA BINOMIAL DISTRIBUTION

$$p(x) = \frac{1}{n+1} \frac{B(a+x, b+n-x)}{B(x+1, n-x+1)B(a,b)} \qquad x = 0,1,2,\ldots,n \qquad a > 0 \qquad b > 0$$

$$\mu = \frac{na}{a+b} \qquad \sigma^2 = \frac{nab(a+b+n)}{(a+b)^2(a+b+1)} \qquad B(a,b) \text{ is the beta function.}$$

## Beta Pascal Distribution

$$p(x) = \frac{\Gamma(x)\Gamma(v)\Gamma(\rho+v)\Gamma(v+x-(\rho+r))}{\Gamma(r)\Gamma(x-r+1)\Gamma(\rho)\Gamma(v-\rho)\Gamma(v+x)} \qquad x = r, r+1, \ldots \quad v > p > 0$$

$$\mu = r\frac{v-1}{\rho-1}, \ \rho > 1 \qquad \sigma^2 = r(r+\rho-1)\frac{(v-1)(v-\rho)}{(\rho-1)^2(\rho-2)}, \ \rho > 2$$

## Binomial Distribution

$$p(x) = \binom{n}{x} p^x q^{n-x} \qquad x = 0,1,2,\ldots,n \quad 0 \le p \le 1 \quad q = 1-p$$

$$\mu = np \qquad \sigma^2 = npq \qquad \beta_1 = \frac{1-2p}{\sqrt{npq}} \qquad \beta_2 = 3 + \frac{1-6pq}{npq}$$

$$M(t) = (q + pe^t)^n \qquad \phi(t) = (q + pe^{it})^n \qquad P(t) = (q + pt)^n$$

## Discrete Weibull Distribution

$$p(x) = (1-p)^{x^\beta} - (1-p)^{(x+1)^\beta} \qquad x = 0,1,\ldots \quad 0 \le p \le 1 \quad \beta > 0$$

## Geometric Distribution

$$p(x) = pq^{1-x} \qquad x = 0,1,2,\ldots \quad 0 \le p \le 1 \quad q = 1-p$$

$$\mu = \frac{1}{p} \qquad \sigma^2 = \frac{q}{p^2} \qquad \beta_1 = \frac{2-p}{\sqrt{q}} \qquad \beta_2 = \frac{p^2 + 6q}{q}$$

$$M(t) = \frac{p}{1-qe^t} \qquad \phi(t) = \frac{p}{1-qe^{it}} \qquad P(t) = \frac{p}{1-qt}$$

## Hypergeometric Distribution

$$p(x) = \frac{\binom{M}{x}\binom{N-M}{n-x}}{\binom{N}{n}} \qquad x = 0,1,2,\ldots,n \quad x \le M \quad n-x \le N-M$$

$$n, M, N, \in N \quad 1 \le n \le N \quad 1 \le M \le N \quad N = 1, 2, \ldots$$

$$\mu = n\frac{M}{N} \quad \sigma^2 = \left(\frac{N-n}{N-1}\right) n \frac{M}{N}\left(1 - \frac{M}{N}\right) \quad \beta_1 = \frac{(N-2M)(N-2n)\sqrt{N-1}}{(N-2)\sqrt{nM(N-M)(N-n)}}$$

$$\beta_2 = \frac{N^2(N-1)}{(N-2)(N-3)nM(N-M)(N-n)}$$

$$\left\{ N(N+1) - 6n(N-n) + 3\frac{M}{N^2}(N-M)\left[N^2(n-2) - Nn^2 + 6n(N-n)\right] \right\}$$

$$M(t) = \frac{(N-M)!(N-n)!}{N!} F(.,e^t) \quad \phi(t) = \frac{(N-M)!(N-n)!}{N!} F(.,e^{it})$$

$$P(t) = \left(\frac{N-M}{N}\right)^n F(.,t)$$

$F(\alpha, \beta, \gamma, x)$ is the hypergeometric function. $\alpha = -n; \quad \beta = -M; \quad \gamma = N - M - n + 1$

## Negative Binomial Distribution

$$p(x) = \binom{x+r-1}{r-1} p^r q^x \quad x = 0, 1, 2, \ldots \quad r = 1, 2, \ldots \quad 0 \le p \le 1 \quad q = 1 - p$$

$$\mu = \frac{rq}{p} \quad \sigma^2 = \frac{rq}{p^2} \quad \beta_1 = \frac{2-p}{\sqrt{rq}} \quad \beta_2 = 3 + \frac{p^2 + 6q}{rq}$$

$$M(t) = \left(\frac{p}{1-qe^t}\right)^r \quad \phi(t) = \left(\frac{p}{1-qe^{it}}\right)^r \quad P(t) = \left(\frac{p}{1-qt}\right)^r$$

## Poisson Distribution

$$p(x) = \frac{e^{-\mu}\mu^x}{x!} \quad x = 0, 1, 2, \ldots \quad \mu > 0$$

$$\mu = \mu \quad \sigma^2 = \mu \quad \beta_1 = \frac{1}{\sqrt{\mu}} \quad \beta_2 = 3 + \frac{1}{\mu}$$

$$M(t) = \exp\left[\mu(e^t - 1)\right] \quad \sigma(t) = \exp\left[\mu(e^{it} - 1)\right] \quad P(t) = \exp\left[\mu(t-1)\right]$$

# Rectangular (Discrete Uniform) Distribution

$$p(x) = 1/n \qquad x = 1, 2, \ldots, n \qquad n \in N$$

$$\mu = \frac{n+1}{2} \qquad \sigma^2 = \frac{n^2 - 1}{12} \qquad \beta_1 = 0 \qquad \beta_2 = \frac{3}{5}\left(3 - \frac{4}{n^2 - 1}\right)$$

$$M(t) = \frac{e^t(1 - e^{nt})}{n(1 - e^t)} \qquad \phi(t) = \frac{e^{it}(1 - e^{nit})}{n(1 - e^{it})} \qquad P(t) = \frac{t(1 - t^n)}{n(1 - t)}$$

# Continuous Distribution Formulas

Probability density function, $f(x)$
Mean, $\mu$
Variance, $\sigma^2$
Coefficient of skewness, $\beta_1$
Coefficient of kurtosis, $\beta_2$
Moment-generating function, $M(t)$
Characteristic function, $\phi(t)$

# Arcsin Distribution

$$f(x) = \frac{1}{\pi\sqrt{x(1-x)}} \qquad 0 < x < 1$$

$$\mu = \frac{1}{2} \qquad \sigma^2 = \frac{1}{8} \qquad \beta_1 = 0 \qquad \beta_2 \frac{3}{2}$$

# Beta Distribution

$$f(x) = \frac{\Gamma(\alpha + \beta)}{\Gamma(\alpha)\Gamma(\beta)} x^{\alpha - 1}(1 - x)^{\beta - 1} \qquad 0 < x < 1 \qquad \alpha, \beta > 0$$

$$\mu = \frac{\alpha}{\alpha + \beta} \qquad \sigma^2 = \frac{\alpha\beta}{(\alpha + \beta)^2(\alpha + \beta + 1)} \qquad \beta_1 = \frac{2(\beta - \alpha)\sqrt{\alpha + \beta + 1}}{\sqrt{\alpha\beta}(\alpha + \beta + 2)}$$

$$\beta_2 = \frac{3(\alpha + \beta + 1)\left[2(\alpha + \beta)^2 + \alpha\beta(\alpha + \beta - 6)\right]}{\alpha\beta(\alpha + \beta + 2)(\alpha + \beta + 3)}$$

## Cauchy Distribution

$$f(x) = \frac{1}{b\pi\left(1 + \left(\frac{x-a}{b}\right)^2\right)} \qquad -\infty < x < \infty \qquad -\infty < a < \infty \qquad b > 0$$

$\mu, \sigma^2, \beta_1, \beta_2, M(t)$ do not exist. $\phi(t) = \exp\left[ait - b|t|\right]$

## Chi Distribution

$$f(x) = \frac{x^{n-1} e^{-x^2/2}}{2^{(n/2)-1} \Gamma(n/2)} \qquad x \geq 0 \qquad n \in N$$

$$\mu = \frac{\Gamma\left(\frac{n+1}{2}\right)}{\Gamma\left(\frac{n}{2}\right)} \qquad \sigma^2 = \frac{\Gamma\left(\frac{n+2}{2}\right)}{\Gamma\left(\frac{n}{2}\right)} - \left[\frac{\Gamma\left(\frac{n+1}{2}\right)}{\Gamma\left(\frac{n}{2}\right)}\right]^2$$

## Chi-Square Distribution

$$f(x) = \frac{e^{-x/2} x^{(v/2)-1}}{2^{v/2} \Gamma(v/2)} \qquad x \geq 0 \qquad v \in N$$

$\mu = v \qquad \sigma^2 = 2v \qquad \beta_1 = 2\sqrt{2/v} \qquad \beta_2 = 3 + \dfrac{12}{v} \qquad M(t) = (1-2t)^{-v/2}, \; t < \dfrac{1}{2}$

$$\phi(t) = (1 - 2it)^{-v/2}$$

## Erlang Distribution

$$f(x) = \frac{1}{\beta^n (n-1)!} x^{n-1} e^{-x/\beta} \qquad x \geq 0 \qquad \beta > 0 \qquad n \in N$$

$\mu = n\beta \qquad \sigma^2 = n\beta^2 \qquad \beta_1 = \dfrac{2}{\sqrt{n}} \qquad \beta_2 = 3 + \dfrac{6}{n}$

$$M(t) = (1 - \beta t)^{-n} \qquad \phi(t) = (1 - \beta it)^{-n}$$

## Statistical Methods for Data Analytics

### EXPONENTIAL DISTRIBUTION

$$f(x) = \lambda e^{-\lambda x} \quad x \geq 0 \quad \lambda > 0$$

$$\mu = \frac{1}{\lambda} \quad \sigma^2 = \frac{1}{\lambda^2} \quad \beta_1 = 2 \quad \beta_2 = 9 \quad M(t) = \frac{\lambda}{\lambda - t}$$

$$\phi(t) = \frac{\lambda}{\lambda - it}$$

### EXTREME-VALUE DISTRIBUTION

$$f(x) = \exp\left[-e^{-(x-\alpha)/\beta}\right] \quad -\infty < x < \infty \quad -\infty < \alpha < \infty \quad \beta > 0$$

$\mu = \alpha + \gamma\beta, \quad \gamma \doteq .5772\ldots$ is Euler's constant $\sigma^2 = \dfrac{\pi^2 \beta^2}{6}.$

$$\beta_1 = 1.29857 \quad \beta_2 = 5.4$$

$$M(t) = e^{\alpha t}\Gamma(1 - \beta t), \quad t < \frac{1}{\beta} \quad \phi(t) = e^{\alpha it}\Gamma(1 - \beta it)$$

### F DISTRIBUTION

$$f(x) \frac{\Gamma[(v_1 + v_2)/2] v_1^{v_1/2} v_2^{v_2/2}}{\Gamma(v_1/2)\Gamma(v_2/2)} x^{(v_1/2)-1}(v_2 + v_1 x)^{-(v_1+v_2)/2}$$

$$x > 0 \quad v_1, v_2 \in N$$

$$\mu = \frac{v_2}{v_2 - 2}, v_2 \geq 3 \quad \sigma^2 = \frac{2v_2^2(v_1 + v_2 - 2)}{v_1(v_2 - 2)^2(v_2 - 4)}, \quad v_2 \geq 5$$

$$\beta_1 = \frac{(2v_1 + v_2 - 2)\sqrt{8(v_2 - 4)}}{\sqrt{v_1(v_2 - 6)}\sqrt{v_1 + v_2 - 2}}, \quad v_2 \geq 7$$

$$\beta_2 = 3 + \frac{12\left[(v_2 - 2)^2(v_2 - 4) + v_1(v_1 + v_2 - 2)(5v_2 - 22)\right]}{v_1(v_2 - 6)(v_2 - 8)(v_1 + v_2 - 2)}, \quad v_2 \geq 9$$

$M(t)$ does not exist. $\phi\left(\dfrac{v_1}{v_2}t\right) = \dfrac{G(v_1, v_2, t)}{B(v_1/2, v_2/2)}$

$B(a,b)$ is the beta function. $G$ is defined by

$$(m+n-2)G(m,n,t) = (m-2)G(m-2,n,t) + 2itG(m,n-2,t), \quad m,n > 2$$

$$mG(m,n,t) = (n-2)G(m+2,n-2,t) - 2itG(m+2,n-4,t), \quad n > 4$$

$$nG(2,n,t) = 2 + 2itG(2,n-2,t), \quad n > 2$$

## GAMMA DISTRIBUTION

$$f(x) = \dfrac{1}{\beta^\alpha \Gamma(\alpha)} x^{\alpha-1} e^{-x/\beta} \qquad x \geq 0 \qquad \alpha, \beta > 0$$

$$\mu = \alpha\beta \qquad \sigma^2 = \alpha\beta^2 \qquad \beta_1 = \dfrac{2}{\sqrt{\alpha}} \qquad \beta_2 = 3\left(1 + \dfrac{2}{\alpha}\right)$$

$$M(t) = (1-\beta t)^{-\alpha} \qquad \phi(t) = (1-\beta it)^{-\alpha}$$

## HALF-NORMAL DISTRIBUTION

$$f(x) = \dfrac{2\theta}{\pi} \exp\left[-\left(\theta^2 x^2 / \pi\right)\right] \qquad x \geq 0 \qquad \theta > 0$$

$$\mu = \dfrac{1}{\theta} \qquad \sigma^2 = \left(\dfrac{\pi - 2}{2}\right)\dfrac{1}{\theta^2} \qquad \beta_1 = \dfrac{4-\pi}{\theta^3} \qquad \beta_2 = \dfrac{3\pi^2 - 4\pi - 12}{4\theta^4}$$

## LAPLACE (DOUBLE EXPONENTIAL) DISTRIBUTION

$$f(x) = \dfrac{1}{2\beta} \exp\left[-\dfrac{|x-\alpha|}{\beta}\right] \qquad -\infty < x < \infty \qquad -\infty < \alpha < \infty \qquad \beta > 0$$

$$\mu = \alpha \qquad \sigma^2 = 2\beta^2 \qquad \beta_1 = 0 \qquad \beta_2 = 6$$

$$M(t) = \dfrac{e^{\alpha t}}{1 - \beta^2 t^2} \qquad \phi(t) = \dfrac{e^{\alpha it}}{1 + \beta^2 t^2}$$

# Statistical Methods for Data Analytics

## LOGISTIC DISTRIBUTION

$$f(x) = \frac{\exp[(x-\alpha)/\beta]}{\beta(1+\exp[(x-\alpha)/\beta])^2}$$

$$-\infty < x < \infty \qquad -\infty < \alpha < \infty \qquad -\infty < \beta < \infty$$

$$\mu = \alpha \qquad \sigma^2 = \frac{\beta^2 \pi^2}{3} \qquad \beta_1 = 0 \qquad \beta_2 = 4.2$$

$$M(t) = e^{\alpha t} \pi \beta t \csc(\pi \beta t) \qquad \phi(t) = e^{\alpha i t} \pi \beta i t \csc(\pi \beta i t)$$

## LOGNORMAL DISTRIBUTION

$$f(x) = \frac{1}{\sqrt{2\pi}\sigma x} \exp\left[-\frac{1}{2\sigma^2}(\ln x - \mu)^2\right]$$

$$x > 0 \qquad -\infty < \mu < \infty \qquad \sigma > 0$$

$$\mu = e^{\mu + \sigma^2/2} \qquad \sigma^2 = e^{2\mu + \sigma^2}\left(e^{\sigma^2} - 1\right)$$

$$\beta_1 = \left(e^{\sigma^2} + 2\right)\left(e^{\sigma^2} - 1\right)^{1/2} \qquad \beta_2 = \left(e^{\sigma^2}\right)^4 + 2\left(e^{\sigma^2}\right)^3 + 3\left(e^{\sigma^2}\right)^2 - 3$$

## NONCENTRAL CHI-SQUARE DISTRIBUTION

$$f(x) = \frac{\exp\left[-\frac{1}{2}(x+\lambda)\right]}{2^{\nu/2}} \sum_{j=0}^{\infty} \frac{x^{(\nu/2)+j-1} \lambda^j}{\Gamma\left(\frac{\nu}{2}+j\right) 2^{2j} j!}$$

$$x > 0 \qquad \lambda > 0 \qquad \nu \in N$$

$$\mu = \nu + \lambda \qquad \sigma^2 = 2(\nu + 2\lambda) \qquad \beta_1 = \frac{\sqrt{8}(\nu + 3\lambda)}{(\nu + 2\lambda)^{3/2}} \qquad \beta_2 = 3 + \frac{12(\nu + 4\lambda)}{(\nu + 2\lambda)^2}$$

$$M(t) = (1-2t)^{-\nu/2} \exp\left[\frac{\lambda t}{1-2t}\right] \qquad \phi(t) = (1-2it)^{-\nu/2} \exp\left[\frac{\lambda i t}{1-2it}\right]$$

## Noncentral F Distribution

$$f(x) = \sum_{i=0}^{\infty} \frac{\Gamma\left(\frac{2i+v_1+v_2}{2}\right)\left(\frac{v_1}{v_2}\right)^{(2i+v_1)/2} x^{(2i+v_1-2)/2} e^{-\lambda/2} \left(\frac{\lambda}{2}\right)}{\Gamma\left(\frac{v_2}{2}\right)\Gamma\left(\frac{2i+v_1}{2}\right) v_1! \left(1+\frac{v_1}{v_2}x\right)^{(2i+v_1+v_2)/2}}$$

$$x > 0 \qquad v_1, v_2 \in N \qquad \lambda > 0$$

$$\mu = \frac{(v_1+\lambda)v_2}{(v_2-2)v_1}, \quad v_2 > 2$$

$$\sigma^2 = \frac{(v_1+\lambda)^2 + 2(v_1+\lambda)v_2^2}{(v_2-2)(v_2-4)v_1^2} - \frac{(v_1+\lambda)^2 v_2^2}{(v_2-2)^2 v_1^2}, \quad v_2 > 4$$

## Noncentral T Distribution

$$f(x) = \frac{v^{v/2}}{\Gamma\left(\frac{v}{2}\right)} \frac{e^{-\delta^2/2}}{\sqrt{\pi}(v+x^2)^{(v+1)/2}} \sum_{i=0}^{\infty} \Gamma\left(\frac{v+i+1}{2}\right)\left(\frac{\delta^i}{i!}\right)\left(\frac{2x^2}{v+x^2}\right)^{i/2}$$

$$-\infty < x < \infty \qquad -\infty < \delta < \infty \qquad v \in N$$

$$\mu_r' = c_r \frac{\Gamma\left(\frac{v-r}{2}\right) v^{r/2}}{2^{r/2}\Gamma\left(\frac{v}{2}\right)}, \quad v > r, \qquad c_{2r-1} = \sum_{i=1}^{r} \frac{(2r-1)!\delta^{2r-1}}{(2i-1)!(r-i)!2^{r-i}},$$

$$c_{2r} = \sum_{i=0}^{r} \frac{(2r)!\delta^{2i}}{(2i)!(r-i)!2^{r-i}}, \qquad r = 1,2,3,\ldots$$

## Normal Distribution

$$f(x) = \frac{1}{\sigma\sqrt{2\pi}} \exp\left[-\frac{(x-\mu)^2}{2\sigma^2}\right]$$

$$-\infty < x < \infty \qquad -\infty < \mu < \infty \qquad \sigma > 0$$

# Statistical Methods for Data Analytics

$$\mu = \mu \qquad \sigma^2 = \sigma^2 \qquad \beta_1 = 0 \qquad \beta_2 = 3 \qquad M(t) = \exp\left[\mu t + \frac{t^2 \sigma^2}{2}\right]$$

$$\phi(t) = \exp\left[\mu i t - \frac{t^2 \sigma^2}{2}\right]$$

## Pareto Distribution

$$f(x) = \theta a^\theta / x^{\theta+1} \qquad x \geq a \qquad \theta > 0 \qquad a > 0$$

$$\mu = \frac{\theta a}{\theta - 1}, \quad \theta > 1 \qquad \sigma^2 = \frac{\theta a^2}{(\theta-1)^2 (\theta-2)}, \qquad \theta > 2$$

$M(t)$ does not exist.

## Rayleigh Distribution

$$f(x) = \frac{x}{\sigma^2} \exp\left[-\frac{x^2}{2\sigma^2}\right] \qquad x \geq 0 \qquad \sigma = 0$$

$$\mu = \sigma\sqrt{\pi/2} \qquad \sigma^2 = 2\sigma^2\left(1 - \frac{\pi}{4}\right) \qquad \beta_1 = \frac{\sqrt{\pi}}{4} \frac{(\pi - 3)}{\left(1 - \frac{\pi}{4}\right)^{3/2}}$$

$$\beta_2 = \frac{2 - \frac{3}{16}\pi^2}{\left(1 - \frac{\pi}{4}\right)^2}$$

## T Distribution

$$f(x) = \frac{1}{\sqrt{\pi v}} \frac{\Gamma\left(\frac{v+1}{2}\right)}{\Gamma\frac{v}{2}} \left(1 + \frac{x^2}{v}\right)^{-(v+1)/2} \qquad -\infty < x < \infty \qquad v \in N$$

$$\mu = 0, \quad v \geq 2 \qquad \sigma^2 = \frac{v}{v-2}, \quad v \geq 3 \qquad \beta_1 = 0, \quad v \geq 4$$

$$\beta_2 = 3 + \frac{6}{v-4}, \quad v \geq 5$$

$M(t)$ does not exist. $\phi(t) = \dfrac{\sqrt{\pi}\,\Gamma\left(\dfrac{v}{2}\right)}{\Gamma\left(\dfrac{v+1}{2}\right)} \displaystyle\int_{-\infty}^{\infty} \dfrac{e^{itz\sqrt{v}}}{(1+z^2)^{(v+1)/2}} dz$

## Triangular Distribution

$$f(x) = \begin{cases} 0 & x \leq a \\ 4(x-a)/(b-a)^2 & a < x \leq (a+b)/2 \\ 4(b-x)/(b-a)^2 & (a+b)/2 < x < b \\ 0 & x \geq b \end{cases}$$

$$-\infty < a < b < \infty$$

$$\mu = \dfrac{a+b}{2} \qquad \sigma^2 = \dfrac{(b-a)^2}{24} \qquad \beta_1 = 0 \qquad \beta_2 = \dfrac{12}{5}$$

$$M(t) = -\dfrac{4\left(e^{at/2} - e^{bt/2}\right)^2}{t^2(b-a)^2} \qquad \phi(t) = \dfrac{4\left(e^{ait/2} - e^{bit/2}\right)^2}{t^2(b-a)^2}$$

## Uniform Distribution

$$f(x) = \dfrac{1}{b-a} \qquad a \leq x \leq b \qquad -\infty < a < b < \infty$$

$$\mu = \dfrac{a+b}{2} \qquad \sigma^2 = \dfrac{(b-a)^2}{12} \qquad \beta_1 = 0 \qquad \beta_2 = \dfrac{9}{5}$$

$$M(t) = \dfrac{e^{bt} - e^{at}}{(b-a)t} \qquad \phi(t) = \dfrac{e^{bit} - e^{ait}}{(b-a)it}$$

## Weibull Distribution

$$f(x) = \dfrac{\alpha}{\beta^\alpha} x^{\alpha-1} e^{-(x/\beta)^\alpha} \qquad x \geq 0 \qquad \alpha, \beta > 0$$

$$\mu = \beta\Gamma\left(1 + \dfrac{1}{\alpha}\right) \qquad \sigma^2 = \beta^2\left[\Gamma\left(1 + \dfrac{2}{\alpha}\right) - \Gamma^2\left(1 + \dfrac{1}{\alpha}\right)\right]$$

$$\beta_1 = \frac{\Gamma\left(1+\frac{3}{\alpha}\right) - 3\Gamma\left(1+\frac{1}{\alpha}\right)\Gamma\left(1+\frac{2}{\alpha}\right) + 2\Gamma^3\left(1+\frac{1}{\alpha}\right)}{\left[\Gamma\left(1+\frac{2}{\alpha}\right) - \Gamma^2\left(1+\frac{1}{\alpha}\right)\right]^{3/2}}$$

$$\beta_2 = \frac{\Gamma\left(1+\frac{4}{\alpha}\right) - 4\Gamma\left(1+\frac{1}{\alpha}\right)\Gamma\left(1+\frac{3}{\alpha}\right) + 6\Gamma^2\left(1+\frac{1}{\alpha}\right)\Gamma\left(1+\frac{2}{\alpha}\right) - 3\Gamma^4\left(1+\frac{1}{\alpha}\right)}{\left[\Gamma\left(1+\frac{2}{\alpha}\right) - \Gamma^2\left(1+\frac{1}{\alpha}\right)\right]^2}$$

## VARIATE GENERATION TECHNIQUES[1]

1. Notation

    Let $h(t)$ and $H(t) = \int_0^t h(\tau)\, d\tau$ be the hazard and cumulative hazard functions, respectively, for a continuous nonnegative random variable $T$, the lifetime of the item under study. The $q \times 1$ vector $z$ contains covariates associated with a particular item or individual. The covariates are linked to the lifetime by the function $\Psi(z)$, which satisfies $\Psi(0) = 1$ and $\Psi(z) \geq 0$ for all $z$. A popular choice is $\Psi(z) = e^{\beta' z}$, where $\beta$ is a $q \times 1$ vector of regression coefficients.

    The cumulative hazard function for $T$ in the *accelerated life* model (Cox and Oakes 1984) is

    $$H(t) = H_0\left(t\, \Psi(z)\right),$$

    where $H_0$ is a baseline cumulative hazard function. Note that when $z = 0$, $H_0 \equiv H$. In this model, the covariates accelerate $(\Psi(z) > 1)$ or decelerate $(\Psi(z) < 1)$, the rate at which the item moves through time. The *proportional* hazards model

    $$H(t) = \Psi(z) H_0(t)$$

    increases $(\Psi(z) > 1)$ or decreases $(\Psi(z) < 1)$ the failure rate of the item by the factor $\Psi(z)$ for all values of $t$.

2. Variate generation algorithms

    The literature shows that the cumulative hazard function, $H(T)$, has a unit exponential distribution. Therefore, a random variate $t$ corresponding to a cumulative hazard function $H(t)$ can be generated by

    $$t = H^{-1}(-\log(u))$$

---

[1] From Leemis, L. M. (1987), "Variate Generation for Accelerated Life and Proportional Hazards Models", *Operations Research*, 35(6), Nov-Dec 1987.

where $u$ is uniformly distributed between 0 and 1. In the accelerated life model, since time is being expanded or contracted by a factor $\Psi(z)$, variates are generated by

$$t = \frac{H_0^{-1}(-\log(u))}{\Psi(z)}$$

In the proportional hazards model, equating $-\log(u)$ to $H(t)$ yields the variate generation formula

$$t = H_0^{-1}\left(\frac{-\log(u)}{\Psi(z)}\right)$$

$$H_{T|T>a}(t) = H(t) - H(a) \quad t > a$$

In the accelerate life model, where $H(t) = H_0(t\Psi(z))$, the time of the next event is generated by

$$t = \frac{H_0^{-1}\left(H_0(a\Psi(z)) - \log(u)\right)}{\Psi(z)}$$

If we equate the conditional cumulative hazard function to $-\log(u)$, the time of the next event in the proportional hazards case is generated by

$$t = H_0^{-1}\left(H_0(a) - \frac{\log(u)}{\Psi(z)}\right)$$

3. Example

$$H(t) = e^{(t/\alpha)^\gamma} - 1 \quad \alpha > 0, \quad \gamma > 0, \quad t > 0$$

and inverse cumulate hazard function

$$H^{-1}(y) = \alpha\left[\log(y+1)\right]^{1/\gamma}$$

Assume that the covariates are linked to survival by the function $\Psi(z) = e^{\beta'z}$ in the accelerated life model. If an NHPP is to be simulated, the baseline hazard function has the exponential power distribution with parameters $\alpha$ and $\gamma$, and the previous event has occurred at time $a$, then the next event is generated at time

$$t = \alpha e^{-\beta'z}\left[\log\left(e^{(\alpha e^{\beta'z}/\alpha)^\gamma} - \log(u)\right)\right]^{1/\gamma},$$

where $u$ is uniformly distributed between 0 and 1.

# REFERENCES

Cox, D. R., and Oakes, D. (1984), *Analysis of Survival Data*, London: Chapman and Hall.

Leemis, L. M. (1987), "Variate Generation for Accelerated Life and Proportional Hazards Models", *Operations Research*, 35(6), 892-894.

# 6 Descriptive Statistics for Data Presentation

*What you see is what the data says.*

## INTRODUCTION

Below is an extensive collection of formulas, equations, models, and templates for descriptive statistics for data presentation. The collection includes many familiar measures as well as not-often-used statistical measures.

## SAMPLE AVERAGE

$$\bar{x} = \frac{1}{n} \sum_{i=1}^{n} x_i$$

*Application areas*: quality control, simulation, facility design, productivity measurement

*Sample calculations*:
Given:
$x_i$: 25, 22, 32, 18, 21, 27, 22, 30, 26, 20
$n = 10$

$$\sum_{i=1}^{10} x_i = 25 + 22 + 32 + 18 + 21 + 27 + 22 + 30 + 26 + 20 = 243$$

$$\bar{x} = 243 / 10 = 24.30$$

## SAMPLE VARIANCE

$$s^2 = \frac{1}{n-1} \sum_{i=1}^{n} (x_i - \bar{x})^2$$

*Application areas*: quality control, simulation, facility design, productivity measurement

The variance and the closely related standard deviation are measures of the extent of the spread of elements in a data distribution. In other words, they are measures of variability in the data set.

*Sample calculations:*
Given:
$x_i$: 25, 22, 32, 18, 21, 27, 22, 30, 26, 20
$n = 10$
$n - 1 = 9$

$$\sum_{i=1}^{10} x_i = 25 + 22 + 32 + 18 + 21 + 27 + 22 + 30 + 26 + 20 = 243$$

$$\bar{x} = 243 / 10 = 24.30$$

$$S^2 = 1/9\{(25 - 24.3)^2 + (22 - 24.3)^2 + K + (32 - 24.3)^2\} = 1/9\{182.10\} = 20.2333$$

*Alternate formulas:*

$$S^2 = \left[\frac{\sum x_i^2 - \frac{(\sum x_i)^2}{n}}{n - 1}\right]$$

$$S^2 = \left[\frac{n(\sum x_i^2) - (\sum x_i)^2}{n(n - 1)}\right]$$

## SAMPLE STANDARD DEVIATION

$$s = \sqrt{s^2}$$

*Application areas*: quality control, simulation, facility design, productivity measurement

The standard deviation formula is simply the square root of the variance. It is the most commonly used measure of spread. An important attribute of the standard deviation as a measure of spread is that if the mean and standard deviation of a normal distribution are known, it is possible to compute the percentile rank associated with any given score. In a normal distribution,

- 68.27% of the data is within one standard deviation of the mean.
- 95.46% of the data is within two standard deviations of the mean.
- 99.73% of the data is within three standard deviations.
- 99.99% of the data is within four standard deviations.
- 99.99985% of the data is within six standard deviations (i.e., within Six Sigma).

# Descriptive Statistics

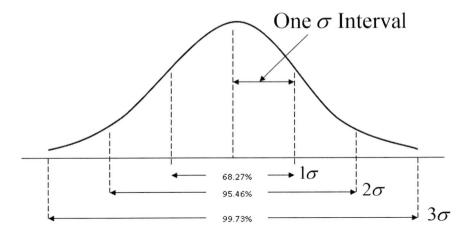

**Note**: A six-sigma (6σ) process is one in which practically **99.99966%** of all products are within quality specs (free of defects).

**FIGURE 6.1** Sigma intervals in normal curve.

Figure 6.1 illustrates the deviation spread of a normal distribution. The standard deviation is used extensively as a measure of spread because it is computationally simple to understand and use. Many formulas in inferential statistics use the standard deviation.

## SAMPLE STANDARD ERROR OF THE MEAN

$$s_m = \frac{s}{\sqrt{n}}$$

*Application areas*: quality control, production planning, packaging, productivity assessment

The standard error of the mean is the standard deviation of the sampling distribution of the mean, where $s$ is the standard deviation of the original distribution and $n$ is the sample size (the number of data points that each mean is based upon). This formula does not assume a normal distribution. However, many of the uses of the formula do assume a normal distribution. The formula shows that the larger the sample size, the smaller the standard error of the mean. In other words, the size of the standard error of the mean is inversely proportional to the square root of the sample size.

## Skewness

$$\text{Skewness} = \frac{n \sum_{i=1}^{n}(x_i - \bar{x})^3}{(n-1)(n-2)s^3}$$

Undefined for $s = 0$ or $n < 3$

## Standardized Skewness

$$\text{Standardized skewness} = \frac{\text{skewness}}{\sqrt{\frac{6}{n}}}$$

## Kurtosis

$$\text{Kurtosis} = \frac{n(n+1)\sum_{i=1}^{n}(x_i - \bar{x})^4}{(n-1)(n-2)(n-3)s^4} - \frac{3(n-1)^2}{(n-2)(n-3)}$$

Undefined for $s = 0$ or $n < 4$

## Standardized Kurtosis

$$\text{Standardized kurtosis} = \frac{\text{Kurtosis}}{\sqrt{\frac{24}{n}}}$$

## Weighted Average

$$\text{Weighted average} = \frac{\sum_{i=1}^{n} x_i w_i}{\sum_{i=1}^{n} w_i}$$

# ESTIMATION AND TESTING

## 100(1 − α)% Confidence Interval for Mean

$$CI = \bar{x} \pm t_{n-1;\alpha/2}\frac{s}{\sqrt{n}}$$

## 100(1 − α)% Confidence Interval for Variance

$$CI = \left[\frac{(n-1)s^2}{\chi^2_{n-1;\alpha/2}}, \frac{(n-1)s^2}{\chi^2_{n-1;1-\alpha/2}}\right]$$

## 100(1 − α)% Confidence Interval for Difference in Means

### For Equal Variance

$$CI = \left(\bar{x}_1 - \bar{x}_2\right) \pm t_{n_1+n_2-2;\alpha/2} s_p \sqrt{\frac{1}{n_1} + \frac{1}{n_2}}$$

# Descriptive Statistics

where

$$s_p = \sqrt{\frac{(n_1-1)s_1^2 + (n_2-1)s_2^2}{n_1 + n_2 - 2}}$$

## For Unequal Variance

$$CI = \left[ (\overline{x}_1 - \overline{x}_2) \pm t_{m;\alpha/2} \sqrt{\frac{s_1^2}{n_1} + \frac{s_2^2}{n_2}} \right]$$

where

$$\frac{1}{m} = \frac{c^2}{n_1 - 1} + \frac{(1-c)^2}{n_2 - 1}$$

and

$$c = \frac{\dfrac{s_1^2}{n_1}}{\dfrac{s_1^2}{n_1} + \dfrac{s_2^2}{n_2}}$$

## $100(1 - \alpha)\%$ Confidence Interval for Ratio of Variances

$$CI = \left(\frac{s_1^2}{s_2^2}\right)\left(\frac{1}{F_{n_1-1,n_2-1;\,\alpha/2}}\right), \left(\frac{s_1^2}{s_2^2}\right)\left(\frac{1}{F_{n_1-1,n_2-1;\,\alpha/2}}\right)$$

## Normal Probability Plot

The input data is first sorted from the smallest to the largest value to compute order statistics. A scatterplot is then generated where the axis positions are computed as follows:

$$\text{Horizontal position} = x_{(i)}$$

$$\text{Vertical position} = \Phi\left(\frac{i - 3/8}{n + 1/4}\right)$$

The labels for the vertical axis are based upon the probability scale using the following expression:

$$100\left(\frac{i - 3/8}{n + 1/4}\right)$$

## COMPARISON OF POISSON RATES

$n_j$ = number of events in sample $j$

$t_j$ = length of sample $j$

Rate estimates: $r_j = \dfrac{n_j}{t_j}$

Rate ratio: $\dfrac{r_1}{r_2}$

Test statistic:

$$z = \max\left(0, \dfrac{\left|n_1 - \dfrac{(n_1+n_2)}{2}\right| - \dfrac{1}{2}}{\sqrt{\dfrac{(n_1+n_2)}{4}}}\right)$$

where $z$ follows the standard normal distribution.

## DISTRIBUTION FUNCTIONS AND PARAMETER ESTIMATION

### BERNOULLI DISTRIBUTION

$$\hat{p} = \bar{x}$$

### BINOMIAL DISTRIBUTION

$$\hat{p} = \dfrac{\bar{x}}{n}$$

where $n$ is the number of trials.

### DISCRETE UNIFORM DISTRIBUTION

$$\hat{a} = \min x_i$$

$$\hat{b} = \max x_i$$

### GEOMETRIC DISTRIBUTION

$$\hat{p} = \dfrac{1}{1+\bar{x}}$$

# Descriptive Statistics

## Negative Binomial Distribution

$$\hat{p} = \frac{k}{\bar{x}}$$

where $k$ is the number of successes.

## Poisson Distribution

$$\hat{\beta} = \bar{x}$$

## Beta Distribution

$$\hat{\alpha} = \bar{x}\left[\frac{\bar{x}(1-\bar{x})}{s^2} - 1\right]$$

$$\hat{\beta} = (1-\bar{x})\left(\frac{\bar{x}(1-\bar{x})}{s^2} - 1\right)$$

## Chi-Square Distribution

If $X_1, \ldots, X_k$ are $k$ independent, normally distributed random variables with mean 0 and variance 1, then the random variable, $Q$, defined as follows, is distributed according to the chi-square distribution with $k$ degrees of freedom.

$$Q = \sum_{i=1}^{k} X_i^2$$

The chi-square distribution is a special case of the gamma distribution and it is represented as:

$$Q \sim \chi_k^2$$

The distribution has one parameter, $k$, which is a positive integer that specifies the number of degrees of freedom (i.e., the number of $X_i$s).

## Erlang Distribution

$$\hat{\alpha} = \text{round}\left(\hat{\alpha} \text{ from Gamma}\right)$$

$$\hat{\beta} = \frac{\hat{\alpha}}{\bar{x}}$$

## Exponential Distribution

$$\hat{\beta} = \frac{1}{\overline{x}}$$

$$\overline{x} = \frac{1}{\hat{\beta}}$$

*Application Areas:*
Common applications of the exponential distribution are for description of the times between events in a Poisson process, in which events occur continuously and independently at a constant average rate, such as queuing analysis and forecasting.

## F Distribution

$$\text{num d.f.: } \hat{v} = \frac{2\hat{w}^3 - 4\hat{w}^2}{\left(s^2(\hat{w}-2)^2(\hat{w}-4)\right) - 2\hat{w}^2}$$

$$\text{den. d.f.: } \hat{w} = \frac{\max(1, 2\overline{x})}{-1 + \overline{x}}$$

## Gamma Distribution

$$R = \log\left(\frac{\text{arithmetic mean}}{\text{geometric mean}}\right)$$

If $0 < R \leq 0.5772$,

$$\hat{\alpha} = R^{-1}\left(0.5000876 + 0.1648852\, R - 0.0544274\, R\right)^2$$

or if $R > 0.5772$,

$$\hat{\alpha} = R^{-1}\left(17.79728 + 11.968477\, R + R^2\right)^{-1}\left(8.898919 + 9.059950\, R + 0.9775373\, R^2\right)$$

$$\hat{\beta} = \hat{\alpha}/\overline{x}$$

## Lognormal Distribution

$$\hat{\mu} = \frac{1}{n}\sum_{i=1}^{n} \log x_i$$

$$\hat{\alpha} = \sqrt{\frac{1}{n-1}\sum_{i=1}^{n}\left(\log x_i - \hat{\mu}\right)^2}$$

# Descriptive Statistics

$$\text{Mean, } \exp\left(\hat{\mu} + \hat{\alpha}^2/2\right)$$

$$\text{Standard deviation, } \sqrt{\exp\left(2\hat{\mu} + \hat{\alpha}^2\right)\left[\exp\left(\hat{\alpha}^2\right) - 1\right]}$$

## Normal Distribution

$$\hat{\mu} = \bar{x}$$

$$\hat{\sigma} = s$$

## Student's t

$$s^2 = \frac{\sum_{i=1}^{n} x_i^2}{n}$$

$$\hat{v} = \frac{2s^2}{-1 + s^2}$$

## Triangular Distribution

$$\hat{a} = \min x_i$$

$$\hat{c} = \max x_i$$

$$\hat{b} = 3\bar{x} - \hat{a} - \hat{c}$$

## Uniform Distribution

$$\hat{a} = \min x_i$$

$$\hat{b} = \max x_i$$

## Weibull Distribution

$$\hat{\alpha} = \frac{n}{\left[\left(\dfrac{1}{\hat{\beta}}\right)^{\hat{a}} \sum_{i=1}^{n} x_i^{\hat{a}} \log x_i - \sum_{i=1}^{n} \log x_i\right]}$$

$$\hat{\beta} = \left(\frac{\sum_{i=1}^{n} x_i^{\hat{\alpha}}}{n}\right)^{1/\hat{\alpha}}$$

## Chi-Square Test for Distribution Fitting

Divide the range of data into non-overlapping classes. The classes are aggregated at each end to ensure that classes have an expected frequency of at least 5.

$O_i$ = observed frequency in class $i$
$E_i$ = expected frequency in class $i$ from fitted distribution
$k$ = number of classes after aggregation

Test statistic

$$x^2 = \sum_{i=1}^{k} \frac{(O_i - E_i)^2}{E_i}$$

follow a chi-square distribution with the degrees of freedom equal to (k-1- # of estimated parameters).

## Kolmogorov–Smirnov Test

$$D_n^+ = \max\left\{\frac{i}{n} - \hat{F}(x_i)\right\}$$

$$1 \leq i \leq n$$

$$D_n^- = \max\left\{\hat{F}(x_i) - \frac{i-1}{n}\right\}$$

$$1 \leq i \leq n$$

$$D_n = \max\left\{D_n^+, D_n^-\right\}$$

where $\hat{F}(x_i)$ = estimated cumulative distribution at $x_i$.

## ANOVA (Analysis of Variance)

### Notations

$k$ = number of treatments
$n_t$ = number of observations for treatment $t$
$\bar{n} = n/k$ = average treatment size

$$n = \sum_{t=1}^{k} n_t$$

$x_{it} = i^{th}$ observation in treatment $t$

$$\bar{x}_t = \text{treatment mean} = \frac{\sum_{i=1}^{n_t} x_{it}}{n_t}$$

# Descriptive Statistics

$$s_t^2 = \text{treatment variance} = \frac{\sum_{i=1}^{n_t}(x_{it}-\overline{x_t})^2}{n_t-1}$$

$$\text{MSE} = \text{mean square error} = \frac{\sum_{t=1}^{k}(n_t-1)s_t^2}{\left(\sum_{t=1}^{k}n_t\right)-k}$$

$$\text{df} = \text{degrees of freedom for the error term} = \left(\sum_{t=1}^{k}n_t\right)-k$$

## Standard Error

$$\sqrt{\frac{s_t^2}{n_t}}$$

Description of Equation: **Standard error (pooled)**
  Formula:

$$\sqrt{\frac{\text{MSE}}{n_t}}$$

## Interval Estimates

$$\overline{x_t} \pm M\sqrt{\frac{\text{MSE}}{n_t}}$$

where
  Confidence interval

$$M = t_{n-k;\alpha/2}$$

Least significant difference (LSD) interval

$$M = \frac{1}{\sqrt{2}}t_{n-k;\alpha/2}$$

## Tukey Interval

$$M = \frac{1}{2}q_{n-k,k;\alpha}$$

where $q_{n-k,k;\alpha}$ = the value of the studentized range distribution with $n-k$ degrees of freedom and $k$ samples such that the cumulative probability equals $1-\alpha$.

## Scheffe Interval

$$M = \frac{\sqrt{k-1}}{\sqrt{2}} \sqrt{F_{k-1,\, n-k;\, \alpha}}$$

## Cochran C-test

This follows $F$ distribution with $\bar{n} - 1$ and $(\bar{n} - 1)(k - 1)$ degrees of freedom.
Test statistic:

$$F = \frac{(k-1)C}{1-C}$$

where

$$C = \frac{\max s_t^2}{\sum_{t=1}^{k} s_t^2}$$

## Bartlett Test

Test statistic:

$$B = 10^{\frac{M}{(n-k)}}$$

$$M = (n-k)\log_{10} \text{MSE} - \sum_{t=1}^{k}(n_t - 1)\log_{10} s_t^2$$

The significance test is based on

$$\frac{M(\ln 10)}{1 + \frac{1}{3(k-1)}\left[\sum_{t=1}^{k}\frac{1}{(n_t - 1)} - \frac{1}{N-k}\right]} \sim \chi^2_{k-1}$$

which follows a chi-square distribution with $k-1$ degrees of freedom.

## Hartley's Test

$$H = \frac{\max(s_t^2)}{\min(s_t^2)}$$

# Descriptive Statistics

## KRUSKAL–WALLIS TEST

Average rank of treatment:

$$\bar{R}_t = \frac{\sum_{i=1}^{n_t} R_{it}}{n_t}$$

If there are no ties, test statistic is

$$w = \left(\frac{12}{n}\sum_{i=1}^{k} n_t \bar{R}_t^{\,2}\right) - 3(n+1)$$

Adjustment for ties:
Let $u_j$ = number of observations tied at any rank for $j = 1,2,3,\ldots,m$ where $m$ = number of unique values in the sample.

$$W = \frac{w}{1 - \dfrac{\sum_{j=1}^{m} u_j^3 - \sum_{j=1}^{m} u_j}{n(n^2-1)}}$$

Significance level: $W$ follows a chi-square distribution with $k - 1$ degrees of freedom.

## FREIDMAN TEST

$X_{it}$ = observation in the $i^{\text{th}}$ row, $t^{\text{th}}$ column

$i = 1,2,\ldots,n \quad t = 1,2,\ldots k$

$R_{it}$ = rank of $X_{it}$ within its row

$n$ = common treatment size (all treatment sizes must be the same for this test)

$$R_t = \sum_{i=1}^{n} R_{it}$$

average rank

$$\bar{R}_t = \frac{\sum_{i=1}^{n_t} R_{it}}{n_t}$$

where data are ranked within each row separately.
Test statistic:

$$Q = \frac{12 S(k-1)}{nk(k^2-1) - \left(\sum u^3 - \sum u\right)}$$

where

$$S = \left(\sum_{t=1}^{k} R_i^2\right) - \frac{n^2 k(k+1)^2}{4}$$

$Q$ follows a chi-square distribution with $k$ degrees of freedom.

## Regression

### Notations

$\mathbf{Y}$ = vector of $n$ observations for the dependent variable
$\mathbf{X}$ = $n$ by $p$ matrix of observations for p independent variables, including any constant term, if any

$$\overline{Y} = \frac{\sum_{i=1}^{n} Y_i}{n}$$

*t-Values*

$$t = \frac{b}{S(b)}$$

*Significance Level*

*t*-Values follow the student's *t* distribution with $n - p$ degrees of freedom.

*Adjusted R-Squared*

$$1 - \left(\frac{n-1}{n-p}\right)(1 - R^2)$$

*Standard Error of Estimate*

$$SE = \sqrt{MSE}$$

*Predicted Values*

$$\hat{\mathbf{Y}} = \mathbf{X}\mathbf{b}$$

*Residuals*

$$\mathbf{e} = \mathbf{Y} - \hat{\mathbf{Y}}$$

*Durbin–Watson Statistic*

$$D = \frac{\sum_{i=1}^{n-1}(e_{i+1} - e_i)^2}{\sum_{i=1}^{n} e_i^2}$$

# Descriptive Statistics

*Mean Absolute Error*

$$\frac{\left(\sum_{i=1}^{n}|e_i|\right)}{n}$$

*Predicted Value*

$$\hat{\mathbf{Y}} = \mathbf{Xb}$$

*Standard Error of Predictions*

$$S(\hat{\mathbf{Y}}) = \sqrt{\text{diagonal elements of } \text{MSE}\left(1 + \mathbf{X}(\mathbf{X'X})^{-1}\mathbf{X'}\right)}$$

Standard error of mean response

$$S(\mathbf{Y'}) = \sqrt{\text{diagonal elements of } \text{MSE}\left(\mathbf{X}(\mathbf{X'X})^{-1}\mathbf{X'}\right)}$$

## Statistical Quality Control

$k$ = number of subgroups
$n_j$ = number of observations in subgroup $j$

$j = 1, 2, \ldots, k$
$x_{ij} = i^{\text{th}}$ observation in subgroup $j$

## Subgroup Statistics

*Subgroup Means*

$$\overline{x_j} = \frac{\sum_{i=1}^{n_j} x_{ij}}{n_j}$$

*Subgroup Standard Deviations*

$$s_j = \sqrt{\frac{\sum_{i=1}^{n_j}(x_{ij} - \overline{x_j})^2}{(n_j - 1)}}$$

*Subgroup Range*

$$R_j = \max\{x_{ij} | 1 \le i \le n_j\} - \min\{x_{ij} | 1 \le i \le n_j\}$$

## X-Bar Charts

$$\overline{\overline{x}} = \frac{\sum_{j=1}^{k} n_i \overline{x_j}}{\sum_{j=1}^{k} n_i}$$

$$\bar{R} = \frac{\left(\sum_{j=1}^{k} n_i R_j\right)}{\sum_{j=1}^{k} n_i}$$

$$s_p = \sqrt{\frac{\sum_{j=1}^{k}(n_j - 1)s_j^2}{\sum_{j=1}^{k}(n_j - 1)}}$$

$$\bar{n} = \frac{1}{k}\sum_{j=1}^{k} n_i$$

Chart based on range:

$$\text{Upper Control Limit: UCL} = \bar{\bar{x}} + A_2 \bar{R}$$

$$\text{Lower Control Limit: LCL} = \bar{\bar{x}} - A_2 \bar{R}$$

Chart based on sigma:

$$\text{UCL} = \bar{\bar{x}} + \frac{3s_p}{\sqrt{n}}$$

$$\text{LCL} = \bar{\bar{x}} - \frac{3s_p}{\sqrt{n}}$$

Chart based on known sigma:

$$\text{UCL} = \bar{\bar{x}} + 3\frac{\sigma}{\sqrt{n}}$$

$$\text{LCL} = \bar{\bar{x}} - 3\frac{\sigma}{\sqrt{n}}$$

**Capability Ratios**

$$C_P = \frac{\text{USL} - \text{LSL}}{6\hat{\alpha}}$$

$$C_R = \frac{1}{C_P}$$

# Descriptive Statistics

$$C_{PK} = \min\left(\frac{\text{USL} - \bar{\bar{x}}}{3\hat{\alpha}}, \frac{\bar{\bar{x}} - \text{LSL}}{3\hat{\alpha}}\right)$$

## R Charts

$$\text{CL} = \bar{R}$$

$$\text{UCL} = D_4 \bar{R}$$

$$\text{LCL} = \text{Max}\left(0, D_3 \bar{R}\right)$$

## S Charts

$$\text{Control Limit: CL} = s_P$$

$$\text{UCL} = s_P \sqrt{\frac{\chi^2_{n-1;\alpha}}{n-1}}$$

$$\text{LCL} = s_P \sqrt{\frac{\chi^2_{n-1;\alpha}}{n-1}}$$

## C Charts

$$\bar{c} = \frac{\sum u_j}{\sum n_j} \qquad \text{UCL} = \bar{c} + 3\sqrt{\bar{c}}$$

$$\text{LCL} = \bar{c} - 3\sqrt{\bar{c}}$$

where $u_j$ = number of defects in the $j$th sample.

## U Charts

$$\bar{u} = \frac{\text{number of defects in all samples}}{\text{number of units in all samples}} = \frac{\sum u_j}{\sum n_j}$$

$$\text{UCL} = \bar{u} + \frac{3\sqrt{\bar{u}}}{\sqrt{n}}$$

$$\text{LCL} = \bar{u} - \frac{3\sqrt{\bar{u}}}{\sqrt{n}}$$

## P Charts

$$p = \frac{\text{number of defective units}}{\text{number of units inspected}}$$

$$\bar{p} = \frac{\text{number of defectives in all samples}}{\text{number of units in all samples}} = \frac{\sum p_j n_j}{\sum n_j}$$

$$\text{UCL} = \bar{p} + \frac{3\sqrt{\bar{p}(1-\bar{p})}}{\sqrt{n}}$$

$$\text{LCL} = \bar{p} - \frac{3\sqrt{\bar{p}(1-\bar{p})}}{\sqrt{n}}$$

## NP Charts

$$\bar{p} = \frac{\sum d_j}{\sum n_j},$$

where $d_j$ is the number of defectives in the $j$th sample.

$$\text{UCL} = \bar{n}\,\bar{p} + 3\sqrt{\bar{n}\,\bar{p}(1-\bar{p})}$$

$$\text{LCL} = \bar{n}\,\bar{p} - 3\sqrt{\bar{n}\,\bar{p}(1-\bar{p})}$$

## CuSum Chart for the Mean

Control mean = $\mu$
Standard deviation = $\alpha$
Difference to detect = $\Delta$
Plot cumulative sums $C_t$ versus $t$ where

$$C_t = \sum_{i=1}^{t}(\bar{x}_i - \mu) \quad \text{for } t = 1,2,\ldots,n$$

The V-mask is located at distance

$$d = \frac{2}{\Delta}\left[\frac{\alpha^2/\bar{n}}{\Delta}\ln\frac{1-\beta}{\alpha/2}\right]$$

in front of the last data point.

$$\text{Angle of mast} = 2\tan^{-1}\frac{\Delta}{2}$$

# Descriptive Statistics

$$\text{Slope of the lines} = \pm \frac{\Delta}{2}$$

## TIME SERIES ANALYSIS

### Notations

$x_t$ or $y_t$ = observation at time $t, t = 1, 2, \ldots, n$

$n$ = number of observations

### Autocorrelation at Lag $k$

$$r_k = \frac{c_k}{c_0}$$

where

$$c_k = \frac{1}{n} \sum_{t=1}^{n-k} (y_t - \bar{y})(y_{t+k} - \bar{y})$$

and

$$\bar{y} = \frac{\left( \sum_{t=1}^{n} y_t \right)}{n}$$

$$\text{Standard error} = \sqrt{\frac{1}{n} \left\{ 1 + 2 \sum_{v=1}^{k-1} r_v^2 \right\}}$$

### Partial Autocorrelation at Lag $k$

$\hat{\theta}_{kk}$ is obtained by solving the following equation:

$$r_j = \hat{\theta}_{k1} r_{j-1} + \hat{\theta}_{k2} r_{j-2} + \ldots + \hat{\theta}_{k(k-1)} r_{j-k+1} + \hat{\theta}_{kk} r_{j-k}$$

$j = 1, 2, \ldots, k$

$$\text{Standard error} = \sqrt{\frac{1}{n}}$$

### Cross-Correlation at Lag $k$

$x$ = input time series
$y$ = output time series

where

$$r_{xy}(k) = \frac{c_{xy}(k)}{s_x s_y}, \quad k = 0, \pm 1, \pm 2, \ldots$$

$$c_{xy}(k) = \begin{cases} \dfrac{1}{n} \sum_{t=1}^{n-k} (x_t - \bar{x})(y_{t+k} - \bar{y}) & k = 0, 1, 2, \ldots \\ \dfrac{1}{n} \sum_{t=1}^{n+k} (x_t - \bar{x})(y_{t-k} - \bar{y}) & k = 0, -1, -2, \ldots \end{cases}$$

and

$$S_x = \sqrt{c_{xx}(0)}$$

$$S_y = \sqrt{c_{yy}(0)}$$

**Box-Cox Computation**

$$yt = \frac{(y + \lambda_2)^{\lambda_1} - 1}{\lambda_1 g^{(\lambda_1 - 1)}} \quad \text{if } \lambda_1 > 0$$

$$yt = g \ln(y + \lambda_2) \quad \text{if } \lambda_1 = 0$$

where $g$ = sample geometric mean $(y + \lambda_2)$.

**Periodogram (Computed Using Fast Fourier Transform)**

If $n$ is odd:

$$I(f_i) = \frac{n}{2}(a_i^2 + b_i^2) \quad i = 1, 2, \ldots, \left[\frac{n-1}{2}\right]$$

where

$$a_i = \frac{2}{n} \sum_{t=1}^{n} t_t \cos 2\pi f_i t$$

$$b_i = \frac{2}{n} \sum_{t=1}^{n} y_t \sin 2\pi f_i t$$

$$f_i = \frac{i}{n}$$

# Descriptive Statistics

If $n$ is even, an additional term is added:

$$I(0.5) = n\left(\frac{1}{n}\sum_{t=1}^{n}(-1)^t Y_t\right)^2$$

## CATEGORICAL ANALYSIS

### Notations

$r$ = number of rows in the table
$c$ = number of columns in the table
$f_{i,j}$ = frequency in position (row $i$, column $j$)
$x_i$ = distinct values of row variable arranged in ascending order; $i = 1,\ldots,r$
$y_j$ = distinct values of column variable arranged in ascending order; $j = 1,\ldots,c$

### Totals

$$R_j = \sum_{j=1}^{c} f_{ij} \qquad C_j = \sum_{i=1}^{r} f_{ij}$$

$$N = \sum_{i=1}^{r}\sum_{j=1}^{c} f_{ij}$$

### Chi-Square

$$\chi^2 = \sum_{i=1}^{r}\sum_{j=1}^{c} \frac{(f_{ij} - E_{ij})^2}{E_{ij}}$$

where

$$E_{ij} = \frac{R_i C_j}{N} \sim \chi^2_{(r-1)(c-1)}$$

### Lambda

$$\lambda = \frac{\left(\sum_{j=1}^{c} f_{max,j} - R_{max}\right)}{N - R_{max}} \quad \text{with rows dependent}$$

$$\lambda = \frac{\left(\sum_{i=1}^{r} f_{i,max} - C_{max}\right)}{N - C_{max}} \quad \text{with columns dependent}$$

$$\lambda = \frac{\left(\sum_{i=1}^{r} f_{i,\max} + \sum_{j=1}^{c} f_{\max,j} - C_{\max} - R_{\max}\right)}{(2N - R_{\max} - C_{\max})} \quad \text{when symmetric}$$

where
  $f_{i\max}$ = largest value in row $i$
  $f_{\max j}$ = largest value in column $j$
  $R_{\max}$ = largest row total
  $C_{\max}$ = largest column total

**Uncertainty Coefficient**

$$U_R = \frac{U(R) + U(C) - U(RC)}{U(R)} \quad \text{with rows dependent}$$

$$U_C = \frac{U(R) + U(C) - U(RC)}{U(C)} \quad \text{with columns dependent}$$

$$U = 2\left(\frac{U(R) + U(C) - U(RC)}{U(R) + U(C)}\right) \quad \text{when symmetric}$$

where

$$U(R) = -\sum_{i=1}^{r} \frac{R_i}{N} \log \frac{R_i}{N}$$

$$U(C) = -\sum_{j=1}^{c} \frac{C_j}{N} \log \frac{C_j}{N}$$

$$U(RC) = -\sum_{i=1}^{r} \sum_{j=1}^{c} \frac{f_{ij}}{N} \log \frac{f_{ij}}{N} \quad \text{for} \quad f_{ij} > 0$$

**Somer's *D* Measure**

$$D_R = \frac{2(P_C - P_D)}{\left(N^2 - \sum_{j=1}^{c} C_j^2\right)} \quad \text{with rows dependent}$$

$$D_C = \frac{2(P_C - P_D)}{\left(N^2 - \sum_{i=1}^{r} R_i^2\right)} \quad \text{with columns dependent}$$

# Descriptive Statistics

$$D = \frac{4(P_C - P_D)}{\left(N^2 - \sum_{i=1}^{r} R_i^2\right) + \left(N^2 - \sum_{j=1}^{c} C_j^2\right)} \quad \text{when symmetric}$$

where the number of concordant pairs is

$$P_C = \sum_{i=1}^{r}\sum_{j=1}^{c} f_{ij} \sum_{h<i}\sum_{k<j} f_{hk}$$

and the number of discordant pairs is

$$P_D = \sum_{i=1}^{r}\sum_{j=1}^{c} f_{ij} \sum_{h<i}\sum_{k>j} f_{hk}$$

**Eta**

$$E_R = \sqrt{1 - \frac{SS_{RN}}{SS_R}} \quad \text{with rows dependent}$$

where the total corrected sum of squares for the rows is

$$SS_R = \sum_{i=1}^{r}\sum_{j-1}^{c} x_i^2 f_{ij} - \frac{\left(\sum_{i=1}^{r}\sum_{j-1}^{c} x_i f_{ij}\right)^2}{N}$$

and the sum of squares of rows within categories of columns is

$$SS_{RN} = \sum_{j=1}^{c} \left( \sum_{i=1}^{r} x_i^2 f_{ij} - \frac{\left(\sum_{i=1}^{r} x_i^2 f_{ij}\right)^2}{C_j} \right)$$

$$E_C = \sqrt{1 - \frac{SS_{CN}}{SS_C}} \quad \text{with columns dependent}$$

where the total corrected sum of squares for the columns is

$$SS_C = \sum_{i=1}^{r}\sum_{j=1}^{c} x_i^2 f_{ij} - \frac{\left(\sum_{i=1}^{r}\sum_{j=1}^{c} y_i f_{ij}\right)^2}{N}$$

and the sum of squares of columns within categories of rows is

$$SS_{CN} = \sum_{i=1}^{r} \left[ \sum_{j=1}^{c} y_i^2 f_{ij} - \frac{\left( \sum_{j=1}^{c} y_j^2 f_{ij} \right)^2}{R_i} \right] j$$

**Contingency Coefficient**

$$C = \sqrt{\frac{\chi^2}{(\chi^2 + N)}}$$

**Cramer's V Measure**

$$V = \sqrt{\frac{\chi^2}{N}} \text{ for } 2 \times 2 \text{ tables}$$

$$V = \sqrt{\frac{\chi^2}{N(m-1)}} \text{ for all others where } m = \min(r, c).$$

**Conditional Gamma**

$$G = \frac{P_C - P_D}{P_C + P_D}$$

**Pearson's r Measure**

$$R = \frac{\sum_{j=1}^{c} \sum_{i=1}^{r} x_i y_j f_{ij} - \frac{\left( \sum_{j=1}^{c} \sum_{i=1}^{r} x_i f_{ij} \right) \left( \sum_{j=1}^{c} \sum_{i=1}^{r} y_i f_{ij} \right)}{N}}{\sqrt{SS_R SS_C}}$$

If $R = 1$, no significance is printed. Otherwise, the one-sided significance is based on

$$t = R \sqrt{\frac{N-2}{1-R^2}}$$

**Kendall's Tau b Measure**

$$\tau = \frac{2(P_C - P_D)}{\sqrt{\left( N^2 - \sum_{i=1}^{r} R_i^2 \right) \left( N^2 - \sum_{j=1}^{c} C_j^2 \right)}}$$

# Descriptive Statistics

**Tau C Measure**
$$\tau_C = \frac{2m(P_C - P_D)}{(m-1)N^2}$$

**Overall Mean**
$$\bar{x} = \frac{n_1\bar{x_1} + n_2\bar{x_2} + n_3\bar{x_3} + \ldots + n_k\bar{x_k}}{n_1 + n_2 + n_3 + \ldots + n_k} = \frac{\sum n\bar{x}}{\sum n}$$

**Chebyshev's Theorem**
$$1 - \frac{1}{k^2}$$

**Permutation**
$$p_r^n = \frac{n!}{(n-r)!}$$

!! = factorial operation:
$$n! = n(n-1)(n-2)\ldots(3)(2)(1)$$

**Combination**
$$c_r^n = \frac{n!}{r!(n-r)!}$$

**Failure**
$$q = 1 - p = \frac{n-s}{n}$$

# 7 Data Analytics Tools for Understanding Random Field Regression Models*

A model is worth a thousand narratives.

## INTRODUCTION

We consider here random field regression (RFR) models in which an observed outcome variable is modeled as the realization of (typically Gaussian) process in many dimensions. Such models are popular choices for modeling data from a number of different areas of application; for example, the well-established kriging approach in geostatistics (Handcock and Stein, 1993; Nychka, 1999) exploits random fields to model the spatial pattern of mineral ore deposits. Kriging has also been applied in fields such as agriculture and hydrology (Cressie, 1991) and environmental monitoring (Federov, 1996). More recently, RFR models have been proposed for use with computer experiments (Sacks et al., 1989; Currin et al., 1991; Welch et al., 1992; Schonlau et al., 1996; Aslett et al., 1998; Chang et al., 1999; Simpson and Mistree, 2001; van Beers and Kleijnen, 2003; Allen et al., 2003). Empirical evidence has shown that RFR models can generate accurate predictions at unobserved sites. However, a drawback is that RFR models are difficult to interpret and to understand, especially in contrast to conventional regression models. In this chapter, we show that there are actually some close ties between random field and conventional regression models, and we develop some simple data analytic tools that can aid in understanding RFR models. We hope that our results will enhance the understanding of RFR models and encourage their use in applications.

We focus here in particular on computer experiments, in which a complex phenomenon is explored by varying the inputs to a computer code that simulate the phenomenon. Such simulations are commonplace in many areas of science and engineering, because the great increase in computing power makes it increasingly cost-effective to replace actual laboratory experimentation with computer code. Our ideas are also valid for geostatistical and environmental applications; however, we think they are most useful for the high-dimensional input spaces that typify computer experiments.

---

* This chapter is a reprint of a 2004 journal publication by David M. Steinberg and Dizza Bursztyn with the following citation:
  Steinberg, David M. and Dizza Bursztyn (2004), "Data Analytic Tools for Understanding Random Field Regression Models," *Technometrics*, 46:4, 411-420, DOI: 10.1198/004017004000000419 (https://doi.org/10.1198/004017004000000419)

Computer experiments have many features in common with physical experiments. The goal is typically to relate one or more output characteristics to the inputs to the code, and it is important to choose an effective design, as reflected in the sets of input conditions to the program. An important difference between computer experiments and physical experiments is that the former have no random variation—a replicate run with the same set of input conditions generates exactly the same output value. The lack of random error in computer experiments has provided some of the motivation for adopting RFR models for data analysis.

We describe the RFR model and some particular versions that have been proposed for use with computer experiments. In Section 3, we describe two representative examples of the use of RFR models. In Section 4, we demonstrate how RFR models are related to Bayesian regression models. We exploit that correspondence in Section 5 to develop some simple data analytic tools that expose the regression model that is hidden away in a random field model. We apply our ideas to several examples, including an experiment with 20 input variables and 50 observations analyzed by Welch et al. (1992). In Section 6, we provide some additional theory to show why the data analytic tools work. In Section 7, we show how our ideas relate to some special RFR models, and in Section 8, we provide final comments and discussion.

## RFR MODELS

RFR models are used to model the relationship between a vector $x$ of input variables and an output $Y(x)$. We limit the discussion here to the case of a single output, though some applications will have many output variables. The general model is

$$Y(x) = \beta_0 + \sum_{s=1}^{k} \beta_s f_s(x) + Z(x) + \varepsilon(x) \tag{7.1}$$

in which the $f_s(x)$ are fixed regression functions, the $\beta_s$ are fixed unknown parameters, $\varepsilon(x)$ is a random error term, and $Z(x)$, the departure from the linear model, is a random field with mean 0 and with covariance function $C(x_1, x_2)$. In computer experiments and some other applications, the random error term is superfluous. In many settings, it is assumed that $Z(x)$ has a constant variance $\sigma^2$ so that $C(x_1, x_2) = \sigma^2 R(x_1, x_2)$, with $R(x_1, x_2)$ the correlation function for the field. The fixed regression terms, with the exception of the constant, are often not included in the model. For example, Welch et al. (1992) stated that including additional regression terms does not, in their experience, lead to better predictors. We explore this point further in Sections 5 and 7.

For these stochastic models, the response function $Y(x)$ is estimated by the best linear unbiased predictor (BLUP) (Robinson, 1991; Currin et al., 1991; Morris et al., 1993). Given a design $D = \{x_1, \ldots, x_n\}$ and data $Y_D = [Y(x_1), \ldots, Y(x_n)]'$, the BLUP of $Y(x)$ is

$$\hat{Y}(x) = f'(x)\hat{\beta} + c'(x)C^{-1}(Y_D - F\hat{\beta}) \tag{7.2}$$

where $\hat{\beta} = (F'C^{-1}F)^{-1}F'C^{-1}Y_D$ is the generalized least squares estimator of $\beta$, and we use the following notation:

$$f(x) = [1, f_1(x), \ldots, f_k(x)]'$$

$$F = \begin{pmatrix} f'(x_1) \\ \vdots \\ f'(x_n) \end{pmatrix}$$

is the $n \times (k+1)$ expanded regression matrix.

$$C = \{C(x_i, x_j)\} \quad 1 \le i \le n \quad 1 \le j \le n$$

is the $n \times n$ covariance matrix of $(Z(x_1), \ldots, Z(x_n))$ and

$$c(x) = [C(x_1, x), \ldots, C(x_n, x)]'$$

is the vector of covariances between the $Z$'s at the design sites and the estimation input $x$.

We make the following remarks on this model:

1. Any parameters in the covariance function can be estimated by maximum likelihood or cross-validation.
2. If constant variance has been assumed, then all of the covariances in the foregoing equations can be replaced by correlations.
3. If there is no random error term, then the BLUP interpolates the observed data, $Y(x_i) = \hat{Y}(x_i)$.
4. If there are random errors with constant variance $\sigma^2$, then the BLUP has the same basic form, but throughout the equations, the matrix $C$ is replaced by the matrix. $M = C + \sigma^2 I$.

The covariance function plays an important role in the estimator. One general form of covariance function that has been used in a number of applications (see Welch et al. 1992; Aslett et al. 1998) has followed the constant variance assumption with a correlation function given by

$$R(x_1, x_2) = \prod \exp\left\{-\lambda_j |x_{1j} - x_{2j}|^{\alpha(j)}\right\} \quad (7.3)$$

where $\lambda_j \ge 0$ are constants related to the importance of the respective factors in predicting the output, and $\alpha(j) \in (0, 2]$ are constants related to the smoothness of the output with respect to the respective input factor.

Other covariance models have also been suggested. Stein (1989) pointed out that the covariance in Eq. (7.3) has a sharp transition from analytical sample paths when $\alpha(j) = 2$ to completely nondifferentiable sample paths when $\alpha(j) < 2$. He suggested an alternative class of correlation functions that do not have this property. Cressie (1986) described several correlation functions that are popular in kriging.

## TWO EXAMPLES

Welch et al. (1992) analyzed data from a simulated experiment with 20 input factors, each defined on the interval [−.5, .5]. The response was determined largely by six of the input factors from the formula $5X_{12}/(1+X_1)+5(X_4-X_{20})^2+X_5+40X_{19}^3-5X_{19}$. The remaining factors made small linear contribution. They used an RFR model with the correlation function in Eq. (7.3). The estimated correlation function parameters are shown in Table 7.1.

Welch et al. (1992) presented a plot of average effects for the six dominant factors computed from the data and the correlation function. The plot showed the quadratic dependence on both $X_4$ and $X_{20}$ and the cubic dependence on $X_{19}$. We apply our methods to this example in Section 5.

Schonlau et al. (1996) described an experiment to study a computer model of a solar collector. The goal was to relate the heat exchange effectiveness to six input factors (inverse wind velocity, slot width, Reynolds number, admittance, plate thickness, and radiative Nusselt number). The engineers who designed the code desired to obtain a parametric model for the relationship. As a first stage, the random field model with correlation function (7.3) was fitted to the data. Plots of the main effects against each of the six input factors were examined and used to suggest nonlinear parametric functions relating the heat exchange effectiveness to each input. An additive model with these parametric functions was then fitted to the data.

## BAYESIAN REGRESSION MODELS AND RANDOM FIELDS

In this section, we develop some relationship between RFR models and Bayesian regression. Consider the model

$$Y(x) = \beta_0 + \Sigma \beta_s f_s(x) + \varepsilon(x) \tag{7.4}$$

Here we have dropped the term $Z(x)$, which was a key component of the random field model. In most applications, we will want to compensate for the lack of this term by adding more regression functions $f_s(x)$ to the model.

---

### TABLE 7.1
**Estimated Correlation Parameters for a 20-Factor, 50-Run Experiment**

| Factor | $\hat{\lambda}$ | $\hat{\alpha}$ |
|---|---|---|
| 1 | .021 | 2.00 |
| 4 | .036 | 2.00 |
| 5 | .000085 | 2.00 |
| 12 | .011 | 2.00 |
| 19 | .0030 | 1.70 |
| 20 | .030 | 2.00 |
| All others | 0 | 0 |

A Bayesian treatment of the regression model now proceeds by assigning prior distributions to the coefficients. We make the prior assumptions

$$E\{\beta_s\} = 0 \tag{7.5a}$$

$$\text{var}(\beta_s) = \sigma^2 \tau_s^2 \tag{7.5b}$$

and the $\beta_s$ are independent. The constant scaling factor $\sigma^2$ in Eq. (7.5b) is the variance of the experimental error, if that term is included in the model. Otherwise, it is an arbitrary constant. Typically, we will want to assign a vague prior distribution to the constant term using the well-known method of letting $\tau_0^2 \to \infty$ (see, e.g., Lindley and Smith, 1972). We may also want to assign vague priors to some other coefficients.

Separate the summation of regression functions into two parts, with all terms that have vague priors in the first sum and all terms that have proper priors in the second sum.

$$Y(x) = \beta_0 + \sum_{s=1}^{k} \beta_s f_s(x) + \sum_{s=k+1}^{K} \beta_s f_s(x) + \varepsilon(x) \tag{7.6}$$

We denote the second sum by $Z(x)$ and observe that it has a probability distribution induced by the prior assumptions about the regression coefficients. For any $x, E\{Z(x)\} = 0$, and for any pair of points $x_1$ and $x_2$, the covariance of the sums is

$$E\{Z(x_1)Z(x_2)\} = C(x_1, x_2) = \sigma^2 \sum_{s=k+1}^{K} \tau_s^2 f_s(x_1) f_s(x_2) \tag{7.7}$$

The Bayesian model thus leads directly to an RFR model. The particular form of the covariance structures is a consequence of the set of regression functions included in the model and the prior variances of the coefficients. Within the Bayesian paradigm, one can even include an infinite number of regression functions, provided that the terms with vague prior variances generate a regression matrix $F$ that has full column rank and that the sum in Eq. (7.7) converges whenever $x_1 = x_2$.

## DATA ANALYSIS: FINDING THE ASSOCIATED REGRESSION MODEL

In this section, we develop a simple data analytical tool that enables us to discover the Bayesian regression model associated with a given RFR model. Essentially, we reverse the flow of ideas in the previous section, in which we took a regression model as our starting point and derived a random field model from it.

Our methods are driven by the following simple idea. Suppose that we take the covariance function in Eq. (7.7) and evaluate it at each pair of data points, generating an $n \times n$ covariance matrix. We can write that matrix in the form

$$C = \sum_{s=k+1}^{K} \tau_s^2 f_s f_s' \tag{7.8}$$

where $f_s = (f_s(x_1), \ldots, f_s(x_n))'$. We then exploit the fact that the expansion in Eq. (8) is similar in form to a spectral decomposition of the covariance matrix.

Our data analysis approach is comprised of three steps:

1. Compute the $n \times n$ covariance matrix for the design points [or the correlation matrix if $Z(x)$ has constant variance].
2. Compute the eigenvalues and eigenvectors of the covariance matrix.
3. For the leading eigenvalues, use plots and regression analysis to find out how the associated eigenvectors are related to the input factors. Each eigenvector has length $n$, corresponding to the $n$ data sites in the design. We treat each eigenvector as if it were a data vector and related it to the input factors.

The method assumes that the covariance function is fully known. In practice, one would first estimate any unknown parameters in the covariance function and then apply the method. Input factors that do not affect the covariance function [e. g., input factors for which $\lambda_j = 0$ in Eq. (7.3)] should not be included in step 3.

We also point out that the correspondence between Eq. (7.7) and the spectral decomposition of the covariance matrix (7.8) is not perfect. First, Eq. (7.7) may include many terms, whereas the spectral decomposition will include exactly $n$. Second, the eigenvectors of the covariance matrix must be orthogonal, whereas the functions in Eq. (7.7) (even, say, the first $n$) will not generally be orthogonal. Moreover, the orthogonality of these vectors depends not only on the functions themselves, but also on the particular choice of design points. Nonetheless, we find that the spectral decomposition of the covariance matrix is often dominated by a small number of eigenvalues and that the leading eigenvectors are related to simple functions of the input factors. (We give further theoretical support for this observation in Section 6.) Thus, investigation of the leading eigenvalues and eigenvectors often provides a great deal of insight into an equivalent Bayesian regression model.

We now present several examples to illustrate the foregoing analysis. In all of the examples, we use the correlation functions in Eq. (7.3) with only a constant term in the fixed-effects regression model. We note that our methods are general and can be applied to any correlation or covariance function. We have applied the methods to the family of correlation functions proposed by Stein (1989) with results similar to those reported here.

### Example 1

We begin with the simple case of a single input factor studied in a 50-run design with one observation located at the midpoint of each of 50 equal-width bins in the interval [−1, 1]. We consider the class of correlation functions in Eq. (7.3) with $\alpha = 2$ and $\lambda = .05$ values that have been reported in empirical studies (Welch et al., 1992). Figure 7.1 is a semilog plot of the eight leading eigenvalues, which decline rapidly, with most of the weight on just the first few eigenvalues. The correlation matrix for this example is a Toeplitz matrix, and the exponential behavior of the eigenvalues is known from theoretical results (see, e.g., Tilli, 1999). Figure 7.2

Data Analytics Tools 217

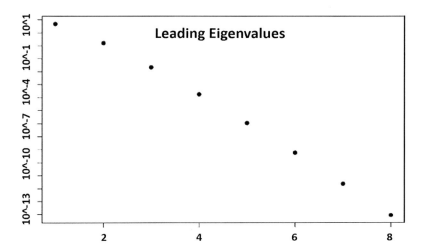

**FIGURE 7.1** The eight leading eigenvalues of the correlation matrix from a single factor (50-run design).

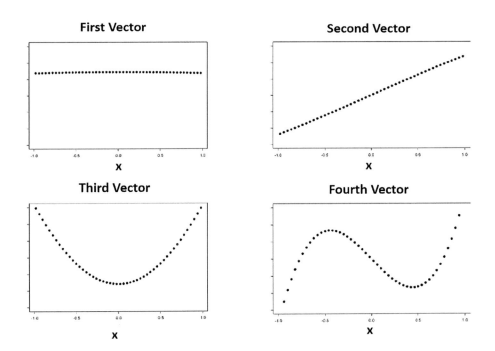

**FIGURE 7.2** The four leading eigenvectors from the correlation matrix of Example 1, as function of the design input $x$.

## TABLE 7.2
### The Six Largest Eigenvalues for a One-Factor, 50-Run Latin Hypercube Experiment for $\lambda = .01, .05, .5, .95$ and for $\alpha = 1, 1.8, 2$

| $\lambda$ | $\alpha$ | 1 | 2 | 3 | 4 | 5 | 6 |
|---|---|---|---|---|---|---|---|
| .01 | 2 | 49.6699 | .3292 | .0009 | 0 | 0 | 0 |
| .05 | 2 | 48.4082 | 1.5705 | .0210 | .0002 | 0 | 0 |
| .50 | 2 | 38.6189 | 9.9307 | 1.3284 | .1144 | .0072 | .004 |
| .95 | 2 | 33.0830 | 13.1859 | 3.1551 | .5087 | .0610 | .0058 |
| .01 | 1.8 | 49.6755 | .3015 | .0140 | .0043 | .0017 | .0009 |
| .05 | 1.8 | 48.4292 | 1.4443 | .0809 | .0219 | .0090 | .0046 |
| .50 | 1.8 | 38.4356 | 9.4297 | 1.5503 | .3160 | .1099 | .0526 |
| .95 | 1.8 | 32.5806 | 12.6858 | 3.3369 | .8153 | .2569 | .1131 |
| .01 | 1 | 49.6686 | .2011 | .0506 | .0226 | .0127 | .0082 |
| .05 | 1 | 48.3764 | .9735 | .2510 | .1124 | .0635 | .0408 |
| .50 | 1 | 36.9457 | 6.9043 | 2.2580 | 1.0699 | .6174 | .4007 |
| .95 | 1 | 29.4151 | 9.6032 | 3.7916 | 1.9107 | 1.1310 | .7434 |

shows the four most important eigenvectors for this case. The leading eigenvector is essentially constant, and the following eigenvectors correspond almost perfectly to linear, quadratic, and cubic polynomials on [−1, 1]. Subsequent eigenvectors continue this pattern. Thus, we conclude that the FRF model, over the range of the data, is roughly equivalent to a Bayesian polynomial regression, with the coefficients of high-degree terms downweighted heavily.

Changing the parameters in the correlation functions (7.3) affects both the rate at which the eigenvalues decline to 0 and the weight on the leading eigenvalues. We repeated our analysis for $\lambda = .01, .05, .5, .95$ and $\alpha = 1, 1.8, 2$. Table 7.2 presents results for the six largest eigenvalues in each case. Our conclusions can be summarized as follows:

1. The leading eigenvectors in all cases are roughly polynomials of increasing degree.
2. As $\lambda$ increase the first eigenvalue decreases and the eigenvalues decay more slowly to 0, which means that the model is more complex, with more weight on higher degree polynomial terms.
3. As $\alpha$ decreases, the eigenvalues decay much more slowly to 0, which means that the model becomes more complex, with more weight on the higher degree terms in the polynomial model.

### Example 2

Now consider an example with four factors, each limited to the interval [−1, 1], with the same $\lambda$ and $\alpha$ for each factor. The design is a 50-run Latin hypercube (McKay et al., 1979). Figure 7.3 is a semilog plot of the eigenvalues when $\alpha = 2$ and $\lambda = .05$. The leading eigenvector is very nearly constant. Plotting the following eigenvectors against the input factors is not very revealing in this example. Some simple regression analysis helps explain why. We fitted regression models,

# Data Analytics Tools

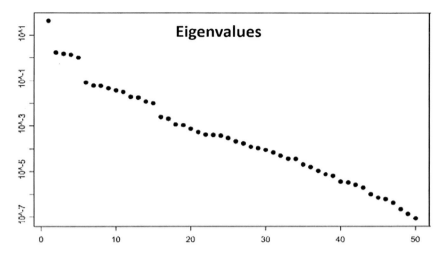

**FIGURE 7.3** A semilog plot of the eigenvalues of the correlation matrix from a four-factor experiment with a 50-run design and a Gaussian correlation function.

## TABLE 7.3
$R^2$ Statistics from the Regressions for v36–v49 (the 15 Leading Eigenvectors, Excluding the First) on the Input Factors

|             | (1)   |       |       | (2)   |       |       | (3)   |       |       |
|-------------|-------|-------|-------|-------|-------|-------|-------|-------|-------|
| Eigenvector | (a)   | (b)   | (c)   | (a)   | (b)   | (c)   | (a)   | (b)   | (c)   |
| 49 | .9992 | .9993 | .9993 | .9567 | .9608 | .9630 | .9876 | .9887 | .9905 |
| 48 | .9992 | .9992 | .9993 | .9527 | .9537 | .9644 | .9825 | .9879 | .9903 |
| 47 | .9994 | .9994 | .9995 | .9543 | .9589 | .9703 | .9802 | .9920 | .9929 |
| 46 | .9992 | .9993 | .9994 | .9417 | .9489 | .9660 | .9051 | .9797 | .9832 |
| 45 | 0 | .1813 | .9995 | .0049 | .1704 | .9734 | .0827 | .9165 | .9228 |
| 44 | 0 | .0382 | .9998 | .0097 | .1047 | .9749 | .0015 | .9370 | .9388 |
| 43 | 0 | .0582 | .9996 | .0132 | .1020 | .9700 | .0138 | .8480 | .8681 |
| 42 | 0 | .2993 | .9996 | .0047 | .2842 | .9717 | .0024 | .8592 | .8637 |
| 41 | 0 | .4518 | .9996 | .0126 | .6065 | .9675 | .0072 | .0506 | .1226 |
| 40 | 0 | .2949 | .9993 | .0059 | .3325 | .9513 | .0080 | .0804 | .2225 |
| 39 | 0 | .5556 | .9989 | .0001 | .2604 | .9542 | .0104 | .0344 | .2122 |
| 38 | 0 | .6780 | .9977 | .0019 | .6470 | .8229 | .0047 | .0741 | .1995 |
| 37 | 0 | .7848 | .9988 | .0093 | .5559 | .7309 | .0019 | .0159 | .2152 |
| 36 | 0 | .6525 | .9983 | .0107 | .3383 | .5400 | .0033 | .0266 | .3248 |

*Note:* The results are given for the following cases: (1) $\lambda = .05$ and $\alpha = 2$, (2) $\lambda = .5$ and $\alpha = 2$, and (3) $\lambda = .05$ and $\alpha = 1$. The $R^2$ statistics are computed for three different models: (a) linear model, (b) linear model with pure quadratics, and (c) full quadratic model.

with each of the four input factors as linear explanatory variables, to the next four eigenvectors. Table 7.3 [column (a), variables 46–49] gives the $R^2$ statistics from those regressions for the following cases: (1) $\lambda = .05$ and $\alpha = 2$; (2) $\lambda = .5$ and $\alpha = 2$;

and (3) $\lambda = .05$ and $\alpha = 1$. All of the $R^2$ statistics are above .9, and most are above .95. For case (1), they are all above .99. The conclusion is that the leading four eigenvectors span almost the same linear space as the linear components of the four factors. Using the same parameter values for each factor leads to a situation where each of these linear terms is "nearly" an eigenvector, but with very similar eigenvalues. As a result, the leading eigenvectors are essentially linear combinations of the linear factor terms.

The subsequent eigenvectors are nearly orthogonal to the linear factor effects and are related to second-order terms, either pure quadratics in the input factors or two-factor interactions. Columns (b) and (c) give the $R^2$ statistics from regressions on linear effects and pure quadratics [column (b)] and a full second-order model [column (c)]. There are ten pure quadratics and two-factor interactions, so we proceed to analyze the next ten eigenvectors. In case (1), we find a perfect correspondence. Indeed, the leading 15 eigenvectors match up almost perfectly to a second-order polynomial with the second-order coefficients downweighted. In case (2), we also find a good fit for seven of the next ten eigenvectors using second-order terms; however, the fit for the next three eigenvectors is less accurate. In both of these cases, comparison of columns (b) and (c) shows that the two-factor interactions have a more prominent role in this model than the pure quadratics. In case (3), with $\alpha = 1$, we find four eigenvectors with good fits from regressions on linear and pure quadratic terms, but adding two-factor interactions is of little value for any of this set of ten eigenvectors. Thus, use of $\alpha = 1$ corresponds to a model with more emphasis on additive complexity in the input factors but less emphasis on interactions. (See also Stein, 1999, section 2.11, which comments that the random field is "locally almost additive" when $\alpha = 1$.)

Our results help explain the finding of Welch et al. (1992) that there is little benefit in adding first-order terms in each factor to their RFR model. Any first-order dependence is effectively included in the random field part of the model, so adding fixed regression terms as well is simply redundant. Moreover, our results show that the constant term is also included in the random field model and so also could be dropped, leaving a model with only the random field part and no fixed regressors.

## Example 3

The use of RFR model is often justified on the grounds that these models provide good predictors. An important issue, raised by an associate editor, is that the prediction function will depend on the data from the simulator, not just on the model parameters. The following example illustrates that the predictor is in fact a combination of the potential of the random field model, which our method explores, and the data.

Suppose that the true response function is a simple low-degree polynomial in the input variable $x_1$. Will the predictor also be a low-degree polynomial as a function of $x_1$? We generated one-factor and three-factor designs (Latin hypercubes) with 20 points and computed the predictor, using the correlation function with $\lambda = .5$ and $\alpha = 2$ for all factors. For the three-factor designs, we assumed no dependence at all on the other factors.

With one factor only, the model potential includes, as seen in Example1, a rich set of low-degree polynomials. We carried out this exercise with polynomials through degree 4 and found that the predictor matched the true function almost perfectly.

Data Analytics Tools

With three factors, the predictor does depend on the model potential that our methods investigate. If the covariance function is estimated, then ideally, the estimation procedure should produce a covariance function that does not depend at all on the two superfluous factors. In that case, we have a one-factor model and the foregoing results again show that the predictor will almost perfectly match the true function. However, if all of the factors are treated symmetrically in the covariance function, the situation is different. Here, as seen in Example 2, the potential model space is dominated by linear terms in each of the three factors (the "real" factor and the two inert factors) and then quadratics and two-factor interactions. When the dependence on $X_1$ is linear, and hence in the model space, the predictor is almost perfectly linear in $X_1$ as well. If the dependence on $X_1$ is cubic, then the predictor still picks up the cubic trend, but plots of $\hat{Y}$ versus $X_1$ for fixed settings of $X_2$ and $X_3$ vary as $X_2$ and $X_3$ are changed. If the dependence is quartic, then the predictor fails to pick up the trend, and plots of $\hat{Y}$ versus $X_1$ vary substantially as $X_2$ and $X_3$ are changed.

### Example 4

We consider here a model with different values of $\lambda$ for each factor and a common value of $\alpha = 2$. We used $\lambda = .01$ for $X_1$, $\lambda = .04$ for $X_2$, $\lambda = .16$ for $X_3$, and $\lambda = .64$ for $X_4$. Thus, $X_4$ should be the most important factor, followed by $X_3$, $X_2$, and $X_1$. The design, as in Example 2, is a 50-run Latin hypercube.

The six largest eigenvalues are e50 = 32.26, e49 = 9.85, e48 = 3.08, e47 = 1.67, e46 = 1.09, and e45 = .88. The leading eigenvector is once again a constant. We fitted first-order regression models for each of the next four eigenvectors. The regression coefficients and $R^2$ statistics are presented in Table 7.4. For v47 and v46, linear regression on the four input factors was not effective. However, adding quadratic terms in $X_3$ and $X_4$ increased $R^2$ to .976 for v47. The model for v46 requires quadratic and cubic terms in $X_3$ and $X_4$, reaching $R^2$ of .983. The higher order terms in $X_4$ and the interaction of $X_3$ with $X_4$ were very important in the explanation of v46 and v47. This example illustrates the crucial role of the scale parameters in the correlation function in the associated regression model. The general form of the correlation function alone does not determine the regression model.

**TABLE 7.4**
**Regression Coefficients and $R^2$ Statistics from a First-Order Regression Model for the Eigenvectors v46–v49, with $\lambda = .01$ for $X_1$, $\lambda = .04$ for $X_2$, $\lambda = .16$ for $X_3$, and $\lambda = .64$ for $X_4$**

| Explained Eigenvector | V46 | V47 | V48 | V49 |
|---|---|---|---|---|
| $R^2$ | .1158 | .0210 | .9648 | .9816 |
| Constant | .0032 | .0182 | .0049 | .0012 |
| $X_1$ Coefficient | .0258 | −.0021 | −.0002 | .0096 |
| $X_2$ Coefficient | .0765 | −.0045 | .0028 | .0045 |
| $X_3$ Coefficient | .0027 | −.0345 | .2395 | -.0118 |
| $X_4$ Coefficient | −.0125 | −.0003 | .0188 | .2407 |

*Note:* $\alpha = 2$ for all of the input variables.

## Example 5

Here we analyze the example from Welch et al. (1992) that we described in Section 3. We generated a 50-run Latin hypercube design for 20 factors (the same type of design use in the original article) and used the estimated correlation function to compute the correlation matrix for our 50 design points. Figure 7.4 shows the 50 eigenvalues on a log scale. The dominant eigenvalue is equal to 49.16 and corresponds to an eigenvector that is almost perfectly constant. Then there is a group of five eigenvalues ranging from .02 to .32. The regression analyses in our step 3 include only the six factors, 1, 4, 5, 12, 19, and 20, that had positive values of $\hat{\lambda}$. The results are summarized in Table 7.5 and show that the group of five eigenvectors is explained perfectly by linear regressions. The largest weight is placed on factors 4 and 20, which have the largest estimated scale parameters and dominate the first two eigenvectors (see Table 7.5). Factor 5, which has by far the smallest value of $\hat{\lambda}$ among the six active factors, plays almost no role at all in the regressions. So these five eigenvectors are equivalent to linear regression on factors 1, 4, 12, 19, and 20.

The remaining eigenvalues are all less than .002. We also modeled the eigenvectors for the largest of these eigenvalues by regression using only factors 1, 4, 12, 19, and 20. First-order regression models produce very poor fits, with $R^2 = .2$, but quadratic regressions have $R^2$ values of .9 and above. The first eigenvector in this group is most closely associated with $X_{19}^2$, $X_1 X_4$, $X_1 X_{20}$, and $X_4^2$. The second eigenvector is most closely related to $X_4 X_{20}$ and $X_{19}^2$, with smaller coefficients for $X_1 X_4$, $X_1 X_{20}$, $X_4^2$, and $X_{20}^2$. A slightly better fit is obtained (with $R^2$ increasing from .90 to .93) by adding a linear effect for factor 5 and a cubic effect for factor 19. (Following the previous example, the low exponent for factor 19 suggests that higher order

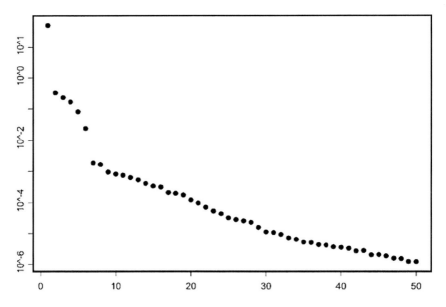

**FIGURE 7.4**  A semilog plot of the eigenvalues of the correlation matrix from the 20-factor experiment.

## TABLE 7.5
## Regression Coefficients and $R^2$ Statistics from a First-Order Regression Model for the Eigenvectors v45–v49 on the Six Active Factors

| Explained Eigenvector | v45 | v46 | v47 | v48 | v49 |
|---|---|---|---|---|---|
| $R^2$ | .999 | 1.000 | 1.000 | 1.000 | 1.000 |
| $x_1$ | .083 | .071 | .484 | .002 | −.083 |
| $x_4$ Coefficient | −.035 | .157 | .124 | .251 | .397 |
| $x_5$ Coefficient | −.003 | 0 | 0 | 0 | 0 |
| $x_{12}$ Coefficient | −.054 | −.504 | .037 | .084 | .003 |
| $x_{19}$ Coefficient | .501 | −.014 | −.011 | .010 | .002 |
| $x_{20}$ Coefficient | −.063 | .197 | −.083 | .422 | −.198 |

*Note:* The constant terms in the regressions were all 0 (to three decimal places) and are not shown.

terms will be most useful for this factor.) The coefficients of these additional terms are similar to those for the four weaker effects listed earlier. The next eigenvector is most closely related to $X_1 X_{20}$, $X_4 X_{20}$, and $X_1 X_4$. Adding a linear effect for factor 5 improves the fit.

The regression shows that further effects of most importance are the pure quadratic effect of factor 19 and the quadratics associated with factors having exponent 2 and large-scale factors (especially factors 4 and 20). Lesser effects are the linear term for factor 5 and the pure cubic for factor 19. Thus, the RFR corresponds to a Bayesian regression model that emphasizes the linear terms in the five main factors with lesser weight to certain higher order terms and the linear effect of factor 5.

The plots of Welch et al. (1992) showed that the fitted model does a good job of picking up the cubic dependence on factor 19 and the quadratic dependence on factors 4 and 20, despite the fact that the model itself places much more weight on the linear effects of these factors.

## RELATING EIGENVECTORS TO REGRESSION FUNCTIONS

In this section, we sketch some theoretical ideas that suggest why the eigenvectors of a covariance function in an RFR model should be related to simple functions of the input factors from an associated Bayesian regression model. In fact, these ideas are essentially a discrete analog of the integral equations that are used to derive the Karhunen–Loève expansion of a stochastic process (see, e. g., Yaglom (1987), section 26.1).

We proceed from the regression model and covariance function in Eqs. (7.6) and (7.7), with the additional assumption that the regression functions $f_s(x)$ are orthonormal with respect to a suitable weight function $w(x)$, that is, $\int f_s(x) f_t(x) w(x) dx = \delta_{s,t}$. Also, we note that the dominant terms in the regression model (i. e., those with large prior variances) will be simple functions like low-degree polynomials. Let $C$ be the $n \times n$ matrix $\left(C(x_i, x_j)\right)_{i,j}$, $i = 1,\ldots,n$, $j = 1,\ldots,n$. We would like to show that simple functions $f_s(x)$ included in the regression model are "approximately" eigenvectors of $C$ in the sense that if $(f_s = f_s(x_1),\ldots,f_s(x_n))'$, then $cf_s \approx \lambda_s f_s$. Note that

$$(Cf_s)_v = \sum_{u=1}^{n} C_{v,u} f_{s,u} = \sum_{u=1}^{n} C(x_v, x_u) f_s(x_u) \tag{7.9}$$

The right side of Eq. (7.9) can be considered a discrete approximation to an integral over the design region with respect to a weight function $w(x)$ that reflects the density from which the design points have been sampled. For Latin hypercube design, a standard recommendation for computer experiments, Owen (1994) derived detailed results on such approximations, with the weight function a uniform density. Good approximation results, again with a uniform density, will also hold for the good lattice point designs proposed by Fang (1980) (see also Fang et al., 2000). Simple functions like low-degree polynomials and low-frequency trigonometric functions have small total variation, and the results of Neidereiter (1992) show that the approximations will then be especially accurate. Applying the integral approximation gives

$$(Cf_s)_v \approx \int C(x_v, x_u) f_s(x_u) w(x_u) dx_u$$

$$= \sigma^2 \int \left( \sum \tau_t^2 f_t(x_v) f_t(x_u) \right) f_s(x_u) w(x_u) dx_u$$

$$= \sigma^2 \sum \tau_t^2 f_t(x_v) \int f_t(x_u) f_s(x_u) w(x_u) dx_u$$

$$= \sigma^2 \tau_t^2 f_s(x_v)$$

$$= \sigma^2 \int \left( \sum \tau_t^2 f_t(x_v) f_t(x_u) \right) f_s(x_u) w(x_u) dx_u$$

$$= \sigma^2 \sum \tau_t^2 f_t(x_v) \int f_t(x_u) f_s(x_u) w(x_u) dx_u$$

$$= \sigma^2 \tau_t^2 f_s(x_v).$$

The last equality derives from the fact that the functions $f_s(x)$ are orthonormal with respect to the weight function $w(x)$.

We thus conclude that the dominant functions in the series expansion (i.e., those with large prior variances), evaluated at the design points, should be nearly parallel to the main eigenvectors of the covariance matrix $C$. This property should hold, in particular, when the dominant functions have small total variation.

Data Analytics Tools                                                                 225

## SOME SPECIAL RANDOM FIELD MODELS

In this section, we explore the analogy between Bayesian regression and RFR models for two special cases. First, we show that the Gaussian correlation function [i.e., Eq. (7.3) in the case where $\alpha = 2$] corresponds exactly to a Bayesian regression model with damped polynomials. Then we show how a Bayesian model for trigonometric regression leads to an RFR model in which the covariance function is a spline. Further expansions similar to those derived here can be found for one-dimensional random fields in Yaglom(1987) and Federov (1996).

## GAUSSIAN COVARIANCE AS DAMPED POLYNOMIAL REGRESSION

The correspondence between the Gaussian RFR model and polynomial regression follows from the results of Steinberg (1985, 1990). Consider first the case where there is a single input factor. Let $H_s(x)$ denote the Hermite polynomial of degree $s$ and define

$$H_s^*(x) = H_s\left(x / \sqrt{2}\right) / \left(2^s s!\right)^{1/2}$$

The sequence of polynomials $H_s^*(x)$ is orthonormal with respect to the standard normal density, so that if $Z \sim N(0,1)$, then

$$E\left\{H_s^*(Z) H_t^*(Z)\right\} = \delta_{s,t}$$

Define a damped version of the polynomials by

$$J_s(x) = H_s^*(x) \exp\left\{-\frac{wx^2}{2(1+w)}\right\}$$

for $0 < W < 1$.

Now assume that the random field can be represented by a series expansion

$$Z(x) = \sum_{s=0}^{\infty} \beta_s J_s(x),$$

in which the higher degree polynomials are downweighted by assuming a priori that $E\{\beta_s\} = 0$, that $\text{var}(\beta_s) = \sigma^2 w^s$ (where $w$ is the same number used in the damping term), and that the $\beta_s$'s are independent. Under these assumptions, $Z(x)$ is a stochastic process with covariance function

$$C(x_1, x_2) = \sigma^2 \sum_{s=0}^{\infty} w^3 J_s(x_1) J_s(x_2)$$

$$= \sigma^2 \left(1 - w^2\right) - 1/2 \exp\left\{-\frac{w}{2(1-w^2)} (x_1 - x_2)^2\right\}$$

The second equality follows from the application of Mehler's formula (Watson, 1933) and is the same as Eq. (7.3) with $\alpha = 2$, $\lambda = \dfrac{w}{2(1-w^2)}$, and an appropriate matching of the variances. Thus, this particular version of the model proposed by Welch et al. (1992) is equivalent to a Bayesian regression model that uses damped polynomials and downweights the terms of high degree. Figure 7.5 presents the plots of $J_1(x)$ and $J_2(x)$ for $\lambda = .2 (w = .3508)$, a case of rather weak damping. Note that within the unit interval, these functions are very similar in form to the second and third eigenvectors in the first example of Section 5.

We now extend these computations to the general case in which there are $p$ input factors $x_1,\ldots,x_p$ and $\alpha(j) = 2$ for all $j$. First, we observe that the correlation function (7.3) for the RFR model in this case is obtained by multiplying the respective one-dimensional correlation functions for each of the factors. Let $R_j(x_{1j}, x_{2j}) = \exp\{-\lambda_j(x_{1j} - x_{2j})^2\}$. Then $R(x_1, x_2) = \prod_{j=1}^{p} R_j(x_{1j}, x_{2j})$. From the preceding result for the one-dimensional case,

$$R_j(x_{1j}, x_{2j}) = \sum_{s=0}^{\infty} w_j^s J_s(x_{1j}) J_s(x_{2j})$$

where $.5 w_j / (1 - w_j^2) = \lambda_j$. Hence,

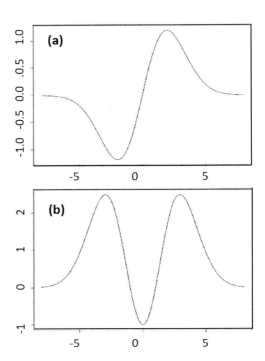

**FIGURE 7.5** Plots of $J_1(x)$ and $J_2(x)$ for $\lambda = .2$ ($w = .3508$).

Data Analytics Tools

$$R(x_1, x_2) = \prod_{j=1}^{p} \sum_{s=0}^{\infty} w_j^s J_s(x_{1j}) J_s(x_{2j})$$

The terms that result from multiplying the $p$ series are damped multivariate polynomials that have the general form

$$\left( \prod_j w_j^{s(j)} \right) J_{s(1)}(x_{11}) J_{s(1)}(x_{21}) \ldots J_{s(p)}(x_{1p}) J_{s(p)}(x_{2p})$$

where $d = \Sigma_j s(j)$ is the degree of the polynomial. Following the same arguments as in the univariate case, we can also obtain this correlation function by adopting a model of the form

$$z(x) = \Sigma \theta_u g_u(x)$$

where the functions $g_u(x)$ have the general form $\prod J_{s(j)}(x_j)$ and make a priori assumptions that $\theta_u \sim N(0, \prod w_j^{s(j)})$ and are independent of one another. (See Steinberg, 1990, for more details on the foregoing model.)

## TRIGONOMETRIC REGRESSION AND SPLINE COVARIANCE

Spline functions have become a popular form of nonparametric regression estimation. (For a detailed account, see Wahba, 1990; for a readable introduction, see Green and Silverman, 1994, especially the first two chapters.) Wahba (1978) showed that splines can be derived as Bayes estimates for a certain prior specification.

We illustrate the link between splines and trigonometric regression by combining the ideas of Wahba (1978) with results of Craven and Wahba (1979) and Steinberg (1985). We present in detail only the case of a single input factor defined on the interval [0, 1].

We begin with some necessary definitions and notation. Our regression model is built from scaled Bernoulli polynomials, $k_j(x) = B_j(x)/j!$. The Bernoulli polynomials are defined recursively on the interval [0, 1], with $B_0(x) = 1$ and subsequent polynomials defined by $dB_{j+1}(x)/dx = (j+1)B_j(x)$ subject to $\int_0^1 B_{j+1}(x) dx = 0$. So, for example, $k_1(x) = x - .5$ and $k_2(x) = .5(x^2 - x + 1/6)$. We also define $k_{2m}(x_1, x_2) = k_{2m}|x_1 - x_2|$.

Craven and Wahba (1979) studied the model $Y(x_i) = g(x_i) + \varepsilon_i$, with $\varepsilon_i \sim N(0, \sigma^2)$ a random error term, and showed that the spline estimate of $g(x)$ of degree $2m - 1$ can be expressed using the scaled Bernoulli polynomials. Scalar multiplication of the covariance terms of Craven and Wahba by $\sigma^2/(n\lambda)$ (in their notation) puts their result in the form of our Eq. (7.2) for data with random errors, $\hat{g}(x) = f'(x)\hat{\beta} + c'(x)M^{-1}(Y - F\hat{\beta})$, where $f(x) = (1, k_1(x), \ldots, k_m(x))$ and $M = C + \sigma^2 I$. The covariance function used to compute the matrix $C$ and the vector

$c(x)$ is given by $C(x_1, x_2) = (-1)^{m-1} k_{2m}(x_1, x_2)/(n\lambda)$. The factor $n\lambda$ serves as a variance ratio, scaling the random field covariance relative to the observational error. The estimator of the fixed regression coefficients is $\hat{\beta} = (F'M^{-1}F + \Delta)^{-1} F'M^{-1}Y$, where $\Delta$ is a matrix of all 0's, except for a 1 in the lower right corner.

The estimator of Craven and Wahba (1979), with the foregoing adjustment, has exactly the form of our Eq. (7.2) except for the matrix $\Delta$. The effect of this matrix is to shrink the coefficient of $k_m(x)$ toward 0 and amount to assigning a proper prior distribution to this coefficient with mean 0 and variance $\sigma^2/(n\lambda)$. Although Craven and Wahba included this term in the fixed-effects part of the model, our representation in Section 4 would have grouped it with the terms that go into defining the covariance kernel of the random field.

We thus obtain a representation for degree $2m - 1$ spline regression in which there are fixed polynomials for all degrees less than $m$ and a covariance kernel of the form $C(x_1, x_2) = [\sigma^2/(n\lambda)][k_m(x_1)k_m(x_2) + (-1)^{m-1}k_{2m}(x_1, x_2)]$. Known results for Bernoulli polynomials (Abramowitz and Stegun, 1964, pp. 804–805) demonstrate that

$$(-1)^{m-1} k_{2m}(x_1, x_2) = 2/(2\pi)^{2m} \sum_{j=1}^{\infty} \left[ \cos(2\pi j x_1)\cos(2\pi j x_2) + \sin(2\pi j x_1)\sin(2\pi j x_2) \right] / j^{2m}$$

Comparing this term with Eq. (7.7), we see that we have a representation of this part of the covariance function in terms of a Bayesian trigonometric regression model.

To summarize the results, the following Bayesian regression model on [0,1] leads to spline regression estimation

$$Y(x) = \beta_0 + \sum_{j=1}^{m} \beta_j k_j(x) + \sum_{s=1}^{\infty} [\beta_{1,s} \cos(2\pi s x) + \beta_{2,s} \sin(2\pi s x)] + \varepsilon(x)$$

with the following prior assumptions. We assign diffuse prior distributions to $\beta_0, \ldots, \beta_{m-1}$. Then we assign proper priors to the remaining coefficients, with mean 0 and variances given by $\text{var}(\beta_m) = \sigma^2/(n\lambda)$ and $\text{var}(\beta_{t,s}) = 2\sigma^2/(2\pi s)^{2m}(n\lambda)$. Assume, a priori, that all the coefficients are independent. The prior distributions depend on $\sigma, \lambda$, and $n$ only via the combined term $\sigma^2/(n\lambda)$.

With more than one factor, the covariance function can be obtained as a product of the univariate functions. This, in turn, corresponds to modeling the response function as a product of univariate trigonometric regression.

## DISCUSSION

We have demonstrated that there are some close links between FRF models for computer experiments and Bayesian regression models. The RFR models that have been proposed for use with computer experiments often have dominant components that are simple regression functions, like low-degree polynomials. Studying the relations to standard regression can help illuminate the results of an RFR model and improve interpretability of the results. We have provided a simple data analysis procedure that helps elucidate those links. The ability to generate simple parametric models that are nearly equivalent to an RFR model can be helpful in promoting understanding of the mechanisms relating the output of a computer experiment to the inputs. This goal led Schonlau et al. (1996) to the idea of using the RFR model to suggest a simple parametric form that would match the main-effects plots from the RFR model. They assumed that an additive model in the input factors would be a reasonable choice. Our work shows that many fitted RFR models also place considerable weight on interactions between factors. Thus, adopting an additive parametric model will not always be an effective substitute for an RFR model.

It is important that we also address a more general issue – namely, what methods of analysis are best suited to data from computer experiments. We think that, in general, it will be necessary to use methods with a great deal of flexibility to automatically scree out unimportant factors and to capture nonlinear relationships. The RFR model has shown in applications that it provides such flexibility. Other approaches to fitting high-dimensional data, like ACE (Breiman and Friedman, 1985) and MARS (Friedman, 1991), may also be useful.

Standard polynomial regression does not provide the same degree of flexibility for discovering key factors and complex relationships. Even with a modest number of input factors, the list of candidate polynomial regressions grows rapidly, and most analyses will not check terms above a rather low degree. Further, polynomials are by nature global functions and cannot capture local behavior as well as methods like the RFR model. So, we think it is wrong to interpret our results as saying that RFR models can be replaced by standard regression. On the contrary, we see our results as a means to provide more insight into RFR models that, we hope, will encourage their use.

Finally, it is important to add that any modeling should attempt to account for the knowledge and understanding of the scientists who are generating the data. Often these scientists will be able to provide a base model that captures the main dependence on the input factors. The data analysis should reflect such knowledge. A natural approach is to use a model with two components like that in our Eq. (7.6). The first term would be a (possibly) nonlinear parametric model reflecting the scientist's intuition, and the second would be a residual dependence term, modeled by a flexible method like RFR. We think it desirable to make the covariance structure for the RFR orthogonal to the parametric model. Orthogonality can be accomplished using the ideas in Section 7 by basing the residual dependence on a series expansion using functions that are orthogonal to the parametric model.

## REFERENCES

Abramowitz, M., and Stegun, I. (1964), *Handbook of Mathematical Functions With Formulas, Graphs and Mathematical Tables, Applied Mathematics Series*, No. 55, U. S. Department of Commerce, National Bureau of Standards, Washington, DC.

Allen, T. T., Bernshteyn, M. A., and Kabiri-Bamoradian, K. (2003), "Constructing metamodels for computer experiments," *Journal of Quality Technology*, 35, 264–274.

Aslett, R., Buck, R. J., Duvall, S. G., Sacks, J., and Welch, W. J. (1998), "Circuit optimization via sequential computer experiments: Design of an output buffer," *Applied Statistics*, 47, 31–48.

Breiman, L., and Friedman, J. H. (1985), "Estimating optimal transformations for multiple-regression and correlation," *Journal of the American Statistical Association*, 80, 580–598.

Chang, P. B., Williams, B. J., Santner, T. J., Notz, W. I., and Bartel, D. L. (1999), "Robust optimization of total joint replacements incorporating environmental variables," *Journal of Biomechanical Engineering*, 121, 304–310.

Craven, P., and Wahba, G. (1979), "Smoothing noisy data with spline funtions," *Numerische Mathematik*, 31, 377–403.

Cressie, N. (1986), "Kriging nonstationary data," *Journal of the American Statistical Association*, 81, 625–634.

Cressie, N. (1991), *Statistics for Spatial Data*, New York: Wiley.

Currin, C., Mitchell, T. J., Morris, M., and Ylvisaker, D. (1991), "Bayesian prediction of deterministic functions, with applications to the design and analysis of computer experiments," *Journal of the American Statistical Association*, 86, 953–963.

Fang, K. T. (1980), "Experimental design by uniform distribution," *Acta Mathematice Appiicatae Sinica*, 3, 363–372.

Fang, K. T., Lin, D. K. J., Winker, P., and Zhang, Y. (2000), "Uniform design: Theory and applications," *Technometrics*, 42, 237–248.

Federov, V. (1996), "Design of Spatial Experiments: Model Fitting and Prediction," in *Handbook of Statistics*, Vol. 13, eds. S. Ghosh and C. R. Rao, Amsterdam: Elsevier, pp. 515–553.

Friedman, J. H. (1991), "Multivariate adaptive regression splines," *The Annals of Statistics*, 19, 1–67.

Handcock, M., and Stein, M. (1993), "A bayesian analysis of kriging," *Technometrics*, 35, 403–410.

Green, P. J., and Silverman, B. W. (1994), *Nonparametric Regression and Generalized Linear Models: A Roughness Penalty Approach*, London: Chapman & Hall.

Lindley, D. V., and Smith, A. F. M. (1972), "Bayes Estimates for the Linear Model" (with discussion), *Journal of the Royal Statistical Society, Ser. B*, 34, 41.

McKay, M. D., Beckman, R. J., and Conover, W. J. (1979), "A comparison of three methods for selecting values of input variables in the analysis of output from a computer code," *Technometrics*, 21, 239–245.

Morris, M. D., Mitchell, T. J., and Ylvisaker, D. (1993), "Bayesian design and analysis of computer experiment: Use of derivatives in surface prediction," *Technometrics*, 35, 243–255.

Neidereiter, H. (1992), *Random Number Generation and Quasi-Monte Carlo Methods*, Philadelphia, PA: SIAM.

Nychka, D. W. (1999), "Spatial process estimates as smoothers," in *Smoothing and Regression. Approaches, Computation and Application*, ed. M. G. Schimek, New York: Wiley, pp. 393–424.

Owen. A. B. (1994), "Controlling correlations in latin hypercube samples," *Journal of the American Statistical Association*, 89, 1517–1522.

Robinson, G. K. (1991), "That BLUP is a good thing: The estimation of random effects," *Statistical Science*, 6, 15–51.

Sacks, J., Schiller, S. B., and Welch, W. J. (1989), "Designs for computer experiments," *Technometrics*, 31, 41–47.

Schonlau, M., Hamada, M., and Welch, W. (1996), "Nonparametric function-fitting to suggest nonlinear parametric models," in *Proceedings of the Section on Physical and Engineering Sciences*, American Statistical Association, Alexandria, VA, pp. 262–267.

Simpson, T. W., and Mistree, F. (2001), "Kriging models for global approximation in simulation-based multidisciplinary design optimization," *AIAA Journal*, 39, 2233–2241.

Stein, M. L. (1989), Comment on "Design and Analysis of Computer Experiments," by Sacks et al., *Statistical Science*, 4, 432–433.

Stein, M. L. (1999), *Statistical Interpolation of Spatial Data: Some theory for Kriging*, New York: Springer-Verlag.

Steinberg, D. M. (1985), "Model Robust Response Surface Designs: Scaling Two-Level Factorials," *Biometrika*, 72, 513–526.

Steinberg, D. M. (1990), "A bayesian approach to flexible modeling of multivariable response functions," *Journal of Multivariate Analysis*, 34, 157–172.

Tilli, P. (1999) "Asymptotic spectral distribution of toeplitz-related matrices," in *Fast Reliable Algorithms for Matrices With Structure*, eds. T. Kaliath and A. H. Sayed, Philadelphia, PA: SIAM.

Van Beers, W. C. M., and Kleijnen, J. P. C. (2003), "Kriging for interpolation in random simulation," *Journal of the Operational Research Society*, 54, 255–262.

Wahba, G. (1978), "Improper priors, spline smoothing and the problem of guarding against model errors in regression," *Journal of the Royal Statistical Society, Ser. B*, 40, 364–372.

Wahba, G. (1990), *Spline Models for Observational Data*, Philadelphia, PA: SIAM.

Watson, G. N. (1933), "Notes on generating functions of polynomials: (2) Hermite Polynomials," *Journal of the London Mathematical Society*, 8, 194–199.

Welch, W. J., Buck, R. J., Sacks, J., Wynn, H. P., Mitchell, T. J., and Morris, M. D. (1992), "Screening, predicting, and computer experiments," *Technometrics*, 34, 15–25.

Yaglom, A. M. (1987), *Correlation Theory of Stationary and Related Random Functions I*, New York: Springer-Verlag.

# 8 Application of DEJI Systems Model to Data Integration

*If there is no integration, there is no implementation.*

## INTRODUCTION TO DATA INTEGRATION

In business and industry, the focus is often solely on the end product. This is okay, provided we recognize the multitude of other factors that can impinge on that end product. The premise of this chapter is that any end product is a function of data quality, data relevance, and data integration. As such, data quality is a requirement for product quality. The Design, Evaluation, Justification, and Integration (DEJI) systems model is presented as a viable technique for achieving data design, data evaluation, data justification, and data integration. The systems technique is a combination of qualitative and quantitative tools.

Organizations collect enormous amounts of data every day, but only a miniscule part of it is organized into a meaningful and useful form. Our overindulgence with data has resulted in the emergence of all sorts of data-centric pursuits nowadays, thereby leading to new areas, such as data mining, big data, data analytics, data science, and data engineering. Data is a means to an end, not an end in itself. We should focus equally strongly on what the data is supposed to achieve concurrently with what the data ought to be. It is through this connectivity or integration that the appropriate data will be collected for the appropriate need with the appropriate level of input quality. The proverbial axiom of "garbage in, garbage out" is, indeed, applicable to the linking of data quality to product quality. Therefore, extra efforts must be directed at improving data quality so as to improve data integration to achieve better product quality. Good data quality is the foundation for good product quality.

As a historical context, I recall the case of late Ross Perot, the billionaire businessman, who contested as an independent candidate for the 1992 and 1996 U.S. presidential election. His campaign was marked by his frequent use of data analytics and data presentation techniques to try and get his points across. While those techniques were effective and welcome in his business world, they did not have much impact for him in the political world. He presidential bid quickly fizzled. He was probably way ahead of his time in terms of using data analytics in 1992. His methodology was not in alignment with what the public wanted at that time. If he were to run today, he might enjoy more traction. The morale of this case example is that data must be integrated with the views and needs of the stakeholders.

## LEVERAGING THE INPUT-CONTROL-OUTPUT-MECHANISM MODEL

The ICOM (Input-Control-Output-Mechanism) model is a good framework for the methodology proposed here. The components of the model are explained as follows:

1. **Inputs.** These are the raw materials that are processed and transformed through some activity (e.g., sheet metal, timber, rubber).
2. **Controls.** Serving as transformation agents, controls provide the influence, direction, guideline, or instruction for how the process is expected to work (e.g., quality standards, customer requirements, benchmarks).
3. **Outputs.:** These are the results of the activity that are ready for transferring to the subsequent processes (e.g., final product, table, chair, widgets).
4. **Mechanisms.** These are the drivers that cause the process to operate (e.g., people, tools, technology, machines).

Figure 8.1 shows an illustration of the ICOM model. A specific application of the ICOM model to a distance learning (DL) educational program is shown in Figure 8.2. Based on the lockdown and social distancing caused by COVID-19, DL became a popular mode of delivering educational contents. Thus, a DL example is aptly relevant here. The inputs to the DL process include data in diverse categories, including the instructor's credentials, the student's academic background, and the academic

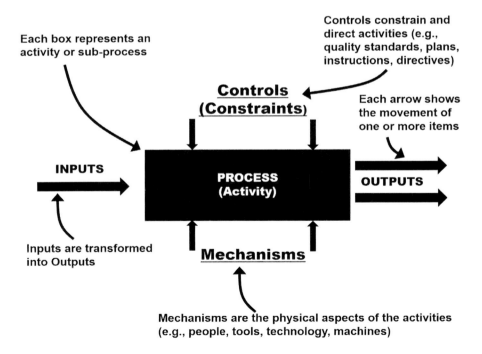

**FIGURE 8.1** ICOM input–process–output framework.

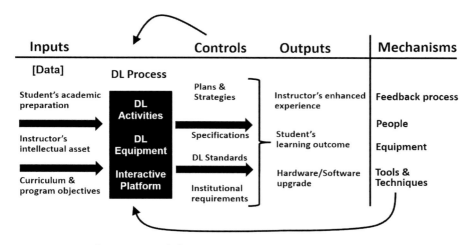

**FIGURE 8.2** ICOM framework applied to DL example.

program objectives. If the fidelity of the input data is high, the potential will be higher for an efficient and effective process. Even though the primary output of the process is the student's learning outcome, there are collateral desirable outputs, such as an enhancement of the instructor's repertoire of expertise and an enhancement of the tools of instruction (e.g., software and hardware upgrades). The mechanisms in this example are people, instructional equipment, and teaching infrastructure (e.g., lecture projection system, online technology). Other application examples can be constructed using the construct of the DL example.

## DATA TYPES AND FIDELITY

There are four primary types of data as summarized in Table 8.1. Every pursuit of quality requires data collection, measurement, and analysis. Data on a nominal scale is the lowest level in the types of measurements. It classifies items into categories. The categories are mutually exclusive and collectively exhaustive; that is, the categories do not overlap and they cover all possible categories of the characteristics being observed. Gender, type of industry, job classification, and color are examples of nominal data. Ordinal scale is distinguished from a nominal scale by the property of order among the categories. We know that first is above second, but we do not know how far above. Similarly, we know that better is preferred to good, but we do not know by how much. In data quality assessment, the A–B–C classification of items based on the Pareto distribution is an example of a measurement on an ordinal scale. The interval data scale is distinguished from an ordinal scale by having equal intervals between the units of measurement. The assignment of priority ratings to quality objectives on a scale of 0–10 is an example of data on an interval scale. Even though a quality factor may have a priority rating of zero, it does not mean that the quality has absolutely no significance to the customer. Similarly, the scoring of zero

**TABLE 8.1**
**Data Types, Characteristics, and Examples**

| Type of data | Data characteristics | Examples |
| --- | --- | --- |
| Nominal data | Classification | Color, Gender, Book type, Attitude |
| Ordinal data | Order | First, Second, Low, High, Good, Better, Rough, Smooth, Happy, Sad |
| Interval | Relative | IQ, Grade Point Average, Temperature, Wealth |
| Ratio | True Zero | Cost, Light Level, Voltage, Floor Space |

on an examination does not imply that a student knows absolutely nothing about the learning objectives. Temperature is a good example of an item that is measured on an interval scale. Even though there is a zero point on the temperature scale, it is an arbitrary relative measure. Other examples of interval scale are IQ measurements and aptitude ratings. Ratio data scale has the same properties of an interval scale, but with a true zero point. For example, an estimate of zero-time unit for the duration of a task is a ratio scale measurement. Other examples of items measured on a ratio scale are cost, time, volume, length, height, weight, inventory level, and number of COVID-19 infections. Many of the data items measured in quality management will be on a ratio scale.

Most quality systems have both quantitative and qualitative data. Quantitative data require that we describe the characteristics of the items being studied numerically. On the other hand, qualitative data are associated with attributes that are not measured numerically. Most items measured on the nominal and ordinal scales will normally be classified into the qualitative data category, whereas those measured on the interval and ratio scales will normally be classified into the quantitative data category. The implication for quality management is that qualitative data can lead to bias in the control mechanism because qualitative data are subject to the personal views and interpretations of the person using the data.

## DATA COLLECTION AND SANITIZATION

For it to serve its purpose as the foundation for product quality, data must be collected, characterized, and sanitized for the intended purpose. It is essential to determine what data to collect for what purposes. Data collection and analysis are the basic components of generating information for production processes to produce acceptable quality of products. The key requirements and best practice for data collection include the following:

- Choosing the data
- Collecting the data
- Doing a relevance check of the data
- Performing a limit check on the data
- Assessing the critical value of the data
- Coding the data for its appropriate use

# Application of DEJI Systems Model

- Processing the data for effective use
- Setting a control limit for data total
- Checking the data for consistency
- Using the appropriate scale of measurement for the data
- Correctly using the information generated from the data

Choosing the data involves selecting data on the basis of their relevance and the level of likelihood that they will be needed for future decisions and whether or not they contribute to making the decision better. The intended users of the data should also be identified. Collecting the data identifies a suitable method of collecting the data as well as the source from which the data will be collected. The collection method will depend on the particular operation being addressed. The common methods include manual tabulation, direct keyboard entry, optical character reader, magnetic coding, electronic scanner, and, more recently, voice command. An input control may be used to confirm the accuracy of collected data. Examples of items to control when collecting data are described as follows:

1. "Relevance Check" to determine if the data is relevant for the scenario in question
2. "Limit Check" to ensure that the data is within known or acceptable limits
3. "Critical Value" to identify boundary points (i.e., control limits) for data values

Coding the data refers to the technique used in representing data in a form useful for generating information. Data quality can be improved if effective data formats and coding are designed into the system right from the beginning. Processing the data is the manipulation of data to generate useful information. Different types of information may be generated from a given data set depending on how it is processed. The processing method should consider how the information will be used, who will be using it, and what caliber of system response time is desired. Processing controls should be used in compliance with the following categories:

1. "Control Total" is used to check for completeness of the processing by comparing accumulated results to a known total
2. "Consistency Check" confirms if the processing is producing the same results for similar data
3. "Scales of Measurement" is used to determine what scale to use for what purpose. For numeric scales, it is desired to specify units of measurement, increments, the zero point on the measurement scale, and the range of values. This is essential for improving data quality.

Using information involves people. Automated systems can collect data, manipulate data, and generate information, but the ultimate usage depends on people. Human decision is initiated by the availability of good data. Even when good data is available, the fidelity of the human using the information correctly may be in question. If the decision is flawed, sometimes we blame the information and the data that generated the information. But the culprit may be the human factors aspect of the decision process. Therein lies human elements in the data–information–decision trifecta. Some of the pertinent human elements include intuition, experience, training, interest, and ethics. The same piece of information that is used positively in one case may be used adversely in another case by the same human.

The timing of data is also very important for its eventual quality potential. The contents, level of detail, and frequency of data can affect the quality process. Data is processed to generate information. Information is analyzed by the decision maker to make the required decisions. Good decisions are based on timely and relevant information, which in turn is based on reliable data. Some factors that are essential for promoting an environment that enables data quality include the following:

- Data summary
- Data processing environment
- Data policies and procedures
- Data momentum: static or dynamic
- Data frequency: often or rarely
- Data constraints
- Data compatibility
- Data contingency

With the foregoing issues, the methodology recommended in this edition of quality insights is the use of a systems framework for the data quality management.

## DEJI SYSTEMS MODEL FOR DATA QUALITY

You don't inspect quality into a product. Rather, you design quality into the data that produces the product. The DEJI model (Badiru, 2014) for systems engineering can be applied to the improvement of data quality. The model provides a structured framework for data design, data evaluation, data justification, and data integration as illustrated in Figure 8.3. Notice that the elements in the steps of the DEJI model overlap to some extent. This is because the stages of the model are not uniquely or brusquely defined. Some elements from each stage can be found in some other elements at the other stages. What is important is to go through the sequential processes of model. The most important aspect is to ensure that the data usage is integrated into the normal and prevailing business scenarios of the organization. A disconnected data set will degenerate to a data set of poor quality.

**Data design.** This involves the concept, requirements, format, logistics, and other desired properties for the data needs in question.

**Data evaluation.** This involves using a combination of qualitative and quantitative tools and techniques to assess the characteristics of the data in question.

**Data justification.** This involves determining why the data may be needed at all. There can be a huge span of data collection, but not all of it may be needed. Justified data requires the investment of time and resources, which must be allocated within the organization's resource allocation process.

# Application of DEJI Systems Model

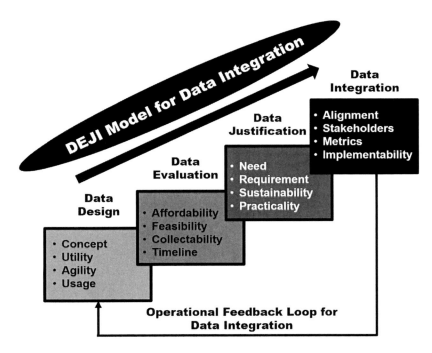

**FIGURE 8.3** Application of DEJI model to data quality management.

**Data integration.** Data utilization must be integrated into the normal and prevailing business practices of the organization. The SMART principle requires data activities that are specific, measurable, aligned, realistic, and timed. If these requirements are not met, the data will be out of sync with the organizational process and may become counterproductive.

## DATA VALUE MODEL

The value of data is proportional to its quality. If quality is improved, the value will be improved. A quantitative assessment of data value can provide an insight into data quality. A technique that is applicable to data value analysis is the project value model (PVM), which is an adaptation of the manufacturing system value (MSV) model presented by Troxler and Blank (1989). The model provides a heuristic decision aid for comparing project alternatives. Value is represented as a deterministic vector function that indicates the value of tangible and intangible attributes that characterize the project. In the case of data quality, we can represent quality as being a function of several attributes as shown below:

$$Q = f(A_1, A_2, \ldots, A_p)$$

where $Q$ = quality, $A = (A_1,\ldots,A_n)$ = vector of quantitative measures or attributes, and $p$ = number of attributes that characterize the data value. Examples of attributes are data classification, throughput, capability, productivity, and cost. Attributes are considered to be a combined function of factors, $x_1$, expressed as

$$A_k(x_1, x_2, \ldots, x_{m_k}) = \sum_{i=1}^{m_k} f_i(x_i)$$

where $\{x_i\}$ = set of $m$ factors associated with attribute $A_k$ ($k = 1, 2, \ldots, p$) and $f_i$ = contribution function of factor $x_i$ to attribute $A_k$. Examples of factors in a data set are data volume, collection reliability, storage flexibility, user fidelity, level of utilization, data security, and data retrieval functionality. Factors are themselves considered to be composed of indicators, $v_i$, expressed as

$$x_i(v_1, v_2, \ldots, v_n) = \sum_{j=1}^{n} z_i(v_i)$$

where $\{v_j\}$ = set of $n$ indicators associated with factor $x_i$ ($i = 1, 2, \ldots, m$) and $z_j$ = scaling function for each indicator variable $v_j$. Examples of indicators are data boundaries, data access, retrieve time requirement, data complexity, and storage size. By combining the above definitions, a composite measure of the data value ($DV$) can, conceptually, be written as follows:

$$DV = f(A_1, A_2, \ldots, A_p)$$

$$= f\left\{ \left[\sum_{i=1}^{m_1} f_i\left(\sum_{j=1}^{n} z_j(v_j)\right)\right]_1, \left[\sum_{i=1}^{m_2} f_i\left(\sum_{j=1}^{n} z_j(v_j)\right)\right]_2, \ldots, \left[\sum_{i=1}^{m_k} f_i\left(\sum_{j=1}^{n} z_j(v_j)\right)\right]_p \right\}$$

where $m$ and $n$ may assume different values for each attribute. A weighting measure to indicate the decision maker's preferences may be included in the model by using an attribute weighting factor, $w_i$, as shown below:

$$PV = f(w_1 A_1, w_2 A_2, \ldots, w_p A_p)$$

where

$$\sum_{k=1}^{p} w_k = 1, \qquad (0 \leq w_k \leq 1)$$

# Application of DEJI Systems Model

In addition to the quantifiable factors, attributes, and indicators that impinge upon the overall DV, the human-based subtle factors should also be included in assessing overall quality of data. Some of such factors are as follows:

- Data communication
- Data cooperation
- Data coordination

## DATA QUALITY CONTROL

Data quality control, in the context of product quality control, refers to the process of regulating or rectifying quality attributes to bring them within acceptable levels. Because of the volatility and dynamism often encountered in complex production systems, it is imperative to embrace the following data quality control practices:

- Recognize humans as drivers of quality.
- Influence the factors that create changes to the quality baseline.
- Ensure there is an agreement for request for data changes.
- Manage the actual changes when and as they occur.
- Monitor performance to detect and understand variances from the baseline.
- Prevent incorrect, inappropriate, or unapproved data changes.
- Use earned value technique (EVT) to track factors affecting data quality.
- Document and disseminate data status.

Human communication or lack thereof can affect data quality and, subsequently, product quality. Communication complexity increases with an increase in the number of communication participants (Badiru, 2008). The statistical formula of combination can be used to estimate the complexity of communication as a function of the number of communication channels or the number of participants involved in data communication, either verbally or in writing. The combination formula is used to calculate the number of possible combinations of $r$ objects from a set of $n$ objects. This is written as follows:

$$_nC_r = \frac{n!}{r![n-r]!}$$

In the case of communication, for illustration purposes, we assume communication is between two members of a team at a time – combination of two from $n$ team members, that is, the number of possible combinations of two members out of a team of $n$ people. Thus, the formula for communication complexity reduces to the expression below, after some of the computation factors cancel out:

$$_nC_2 = \frac{n(n-1)}{2}$$

In a similar vein, we can develop a formula for cooperation complexity based on the statistical concept of permutation. Permutation is the number of possible arrangements of $k$ objects taken from a set of $n$ objects. The permutation formula is written as follows:

$$_nP_k = \frac{n!}{(n-k)!}$$

Thus, the number of possible permutations of two members out of a team of $n$ members is estimated as follows:

$$_nP_2 = n(n-1)$$

Permutation formula is used for cooperation because cooperation is bidirectional. Full cooperation requires that if A cooperates with B, then B must cooperate with A. But A cooperating with B does not necessarily imply B cooperating with A. In notational form,

$$A \to B \text{ does not necessarily imply } B \to A$$

This chapter highlights how data quality ties into product quality. Data is the "raw material" on the basis of which decisions are made relative to the desired product quality. It is best to view data from a structured systems approach, with a desired end goal of data integration. The chapter recommends the use of the DEJI systems model to achieve a structured management of data integration through the stages of design, evaluation, justification, and integration.

## REFERENCES

Badiru, A. B. (2014), "Quality Insights: The DEJI Model for Quality Design, Evaluation, Justification, and Integration," *International Journal of Quality Engineering and Technology*, 4(4), 369–378.

Badiru, A. B. (2008) *Triple C Model of Project Management: Communication, Cooperation, and Coordination*, Boca Raton, FL: Taylor & Francis Group/CRC Press.

Troxler, J. W., and Blank, L. (1989), "A comprehensive methodology for manufacturing system evaluation and comparison," *Journal of Manufacturing Systems*, 8(3), 176–183.

# Index

algebra 83, 127
angle $\phi$ between two lines with direction cosines 108
angle $\psi$ between two lines 103
ANOVA (analysis of variance) 150, 194
application of DEJI systems model to data integration 233
Arcsin distribution 134, 173
area of triangle with vertices 104
areas under a normal curve 187
arithmetic mean 86
autocorrelation at Lag $k$ 162, 203
average 142
average deviation 63
average revenue 59

background in predictive analytics 3
Bardeen–Cooper–Schrieffer equation 15
Bartlett test 152, 196
basic mathematical calculations for data analytics 69
basic probability principles 77, 169
bayesian regression models and random fields 214
Bernoulli 146
Bernoulli distribution 129, 170, 190
beta 146
beta binomial distribution 129, 170
beta distribution 135, 173, 191
beta Pascal distribution 132, 171
binomial 146
binomial distribution 73, 132, 171, 190
Boltzmann's equation 13
box-cox 163
box-cox computation 204

$C$ Charts 160, 201
calculation for data analytics 69
calculus reference for data analytics 32
capability ratios 159, 200
cardioid 105
case example of covidvisualizer website 51
cassinian curves 105
categorical analysis 164, 205
catenary hyperbolic cosine 105
Cauchy distribution 135, 174
Ceva's theorem 93
Chebyshev's theorem 70, 209
chi distribution 135, 174
chi-square 147, 165, 205
chi-square distribution 74, 135, 174, 194
chi-square test for distribution fitting 149, 194

choosing the data 53
circle 105
circular cylinder 100
circular torus 102
Cochran C-Test 151, 196
coding the data 54
collecting the data 53
combination 209
Combinations 70
common constants 20
communication 40
comparison of poisson rates 145, 190
compound functions 34
Conditional Gamma 168, 208
confidence interval for mean 144
confidence interval for ratio of variances 145
confidence interval for variance 144, 188
conflict resolution in data analytics 47
consistency check 55
contingency coefficient 167, 208
continued fractions 114
continuous distributions 134
continuous distribution formulas 173
control total 54
cooperation 45
coordination 47
cosecant curve 106
cosine curve 106
cosine law 127
cotangent curve 106
COVID-19 data analytics 1
Covidvisualize website 51
Cramer's $V$ Measure 168, 208
critical value 54
cross-correlation at Lag $k$ 163, 203
cube 98
cubic equation 88
cubical parabola 106, 107
cumulative distribution function 73
Cusum chart for the mean 161, 202
cyclic quadrilateral 95
cyclic-inscriptable quadrilateral 96
cylinder of cross-sectional area 100
cylindrical coordinates 111

damped polynomial regression 225
data analysis 215
data analytics tools 211
data analytics 69
data and measurements for data analytics 7
data collection and sanitization 236

243

data determination 53
data determination and collection 53
data exploitation 57
data fanaticism 6
data integration 233, 239
data measurement 16, 17
data measurement and comparison 16
data measurement scales 8, 17
data modeling approaches 5
data presentation 185
data processing 54
data quality 238
data quality control 241
data sanitation 236
data scales 17
data types and fidelity 235
data value model 239
data visualization 51
data visualization methods 51
DeMorgan's laws 76
definition of general powers 116
definition of set and notation 74
DEJI model application 233
DEJI systems model 37
DEJI systems model for data integration 233
DEJI systems model for data quality 238
descriptive statistics 185
difference in means 188
Dirac equation 12
direct comparison 16
discrete distributions 129
discrete distribution formulas 170
discrete uniform 146
discrete uniform distribution 173, 190
discrete Weibull distribution 132, 171
discussion 228
distance between two points 102
distance from a point 103
distance from a point to a plane 109
distribution fitting 149
distribution function 71
distribution functions and parameter estimation 146, 190
distribution parameters 142
double exponential distribution 176
Durand's rule 97
dynamism and volatility of data 52

Einstein's equation 11
ellipse 106
ellipsoid 102, 107
elliptic cone 107
elliptic cylinder 107
elliptic paraboloid 108
empirical model building 29
equal variance 144, 188
equation of a line 103

equation of line joining two points 103
equation of plane in intercept form 109
equation of plane passing through points 109
equations of line in parametric form 108
equations of line in standard form 108
equations of line perpendicular to a plane 109
equilateral triangle 92
Erlang 147
Erlang distribution 136, 174, 191
essentials of data analytics 1
estimation 188
estimation and testing 144, 188
Eta 167, 207
expanding 127
expected value 72
exponential 147
exponential distribution 136, 175, 192
extreme-value distribution 136, 175

$F$ distribution 136, 147, 175, 192
factoring 127
factors of higher degree 91
failure 71, 209
fast fourier transform 164, 204
fermat's last theorem 15
finding the associated regression model 215
fisher's exact test 165
formulas from plane analytic geometry 102
Freidman test 153, 197
Frustum of Right Circular Cone 101
fundamental properties 115
fundamental scientific equations 11

gamma distribution 106, 137, 147, 176, 192
Gaussian covariance 225
general equation of a line 103
general equation of a plane 109
general quadrilateral 94
general terms 90
general triangle 92
generalized mean 86
geometric 146
geometric distribution 132, 171, 190
geometric mean 86
geometry 91
global growth of data analytics 2
greek alphabet 83

half-normal distribution 137, 176
harmonic mean 86
Hartley's test 152, 196
Hawking equation 14
Heisenberg's principle 12
Heisenberg's uncertainty principle 12
hyperbolic functions 106
hyperbolic paraboloid 108
hyperboloid of one sheet 108

# Index

hyperboloid of two sheets 108
hypergeometric distribution 133, 171

ICOM model 234
indirect comparison 17
inequalities 113
input-control-output-mechanism model 234
integration by parts 34, 25
integration rules 33
interval estimates 151, 195
Interval Scale of Measurement 18
Interval scale 18
InverseCosineCurve 107
InverseSineCurve 107
InverseTangentCurve 107

Josephson effect 15 Kendall's Taub 168

Kendall's Taub measure 208
Kolmogorov–Smirnov test 149, 194
Kruskal–Wallis test 152, 196
Kurtosis 143, 188

Lagrangian 14
Lambda 165, 205
laplace (double exponential) distribution 138, 176
law of exponents 128
laws of algebraic operations 83
limit check 54
limiting values 113
logarithmic and exponential functions 116
logarithmic curve 107
logarithmici dentities 112
logarithms 128
logarithms to general base 112
logistic distribution 138, 177
lognormal 148
lognormal distribution 138, 177, 192
Lune 101

mathematical equations 21
mathematical signs and symbols 81
Maxwell's equation 13
mean of a binomial distribution 73
mean value $x$ or expected value $\mu$ 78, 169
measurement 7
median revenue 61
Menelaus's theorem 93
methods for data measurement and comparison 16
model environment 29
multivariate control charts 162

Navier-Stokes equation 14
negative binomial distribution 133, 146, 172, 191
nominal scale of measurement 18
non central chi-square distribution 139, 177

non central $F$ distribution 139, 178
non central $t$ distribution 139, 178
non linear regression 156
non-repeated linear factors 89
normal curve 187
normal distribution 73, 140, 178, 193
normal form for equation of a line 103
normal form for equation of plane 109
normal probability plot 145, 189
normal 148
notations 150
NP charts 161, 202
numeric data representation 20

oblate spheroid 102
operations on sets 76
ordinal scale of measurement 18
overall mean 70, 209

$P$ Charts 160, 201
parabola 107
parallelogram 94
parameter estimation 146
pare to distribution 140, 179
partial auto correlation at Lagk 163, 203
partial fractions 89
Pearson'sr 168
Pearson'sr measure 208
periodogram (Computed Using Fast Fourier Transform) 164, 204
permutation 70, 209
planar areas by approximation 97
Planck's black body radiation formula 14
Planck-Einstein equation 13
plane analytic geometry 102
plane curves 105
poisson distribution 73, 134, 172, 191
polar coordinates 105
polynomial approximations 114, 117
population mean 73
powers and roots 85
predictions 155
prism 98
prismatoid 99
probability 71
probability distribution 71
probability terminology 77, 168
prolate spheroid 102
Prolemy's theorem 96
proportion 85
pyramid 98

quadratic equation 69
quadrilaterals 93
quality control 157
quartiles and percentiles 62
quick reference for mathematical equations 21

# Index

R Charts 159, 201
random field(s) 214
random field models 225
random variable 77, 169
range of revenue 63
ratio of variances 189
ratio scale 18
ratio scale measurement 18
raw data 57
Rayleigh distribution 140, 179
rectangle 93
rectangular (discrete uniform) distribution 134, 173
rectangular parallelepiped 98
reference units of measurements 19
regression functions 223
regression models 211
regression statistics 154
regular polygon 95
regular polyhedra 99
relating eigenvectors to regression functions 223
relevance check 54
repeated linear factors 90
RFR models 212
rhombus 94
ridge regression 157
Riemann integral 34
right circular cone 101
right circular cylinder 100
right triangle 92
roots of quadratic 128

S Charts 160, 201
sample average 185
sample standard deviation 186
sample standard error 187
sample variance 64, 185
scales of measurement 55
scheffe interval 151, 196
Schrödinger Equation 12
series expansions 78, 113
set terms and symbols 75
Simpson's rule (n even) 97
Sine law 127
skewness 143, 187
slopes 125
slope M of line joining two points 103
solids bounded by planes 98
solution of cubic equations 87
solution of quadratic equations 87, 89
solving integrals with variable substitution 34
Somer's D measure 166, 206
special product sand factors 83
special values 112
sphere 108
sphere of radius R 100
spherical cap of radius R 101
spherical coordinates 111
spherical sector 101

spherical triangle and polygon 101
spheroids 102
spline covariance 227
standard deviation 65, 143
standard error 143, 195
standard error (internal) 151
standard error (pooled) 151
standard error of the mean 74
standardized kurtosis 143, 188
standardized skewness 143, 188
state-space modeling 30
statistical methods for data analytics 129
statistical quality control 199
statistics for data presentation 185
student's $t$ 148, 193
sub group statistics 157, 199
super conductivity 15
system displays 148
systems modeling 36
systems modeling for data analytics 36
systems view of data analytics 2

tangent curve 107
Tau C 168, 209
$t$ distribution 74, 141, 179
testing 188
the language of data analytics 21
the mode 63
time series analysis 162, 203
trapezoid 94, 97
trapezoidal rule 97
triangles 91
triangular distribution 141, 180, 193
trigonometric ratios 125
trigonometric regression 227
trigonometric solution of the cubic equation 88
triple C questions 39
Tukey Interval 151, 195
two examples 214

U Charts 160, 201
uncertainty coefficient 165, 206
unequal variance 144, 189
uniform distribution 142, 180, 193
units of measurements 19
using the information 55

variance 72, 73, 142
variate generation techniques 181
Venn diagram 75
volatility of data 52

Weddle's rule 98
Weibull distribution 142, 180, 193
weighted average 143, 188

X-Bar Charts 158, 199

zone and segment of two bases 101